Palgrave Studies in the History of Science and Technology

Series Editors

Designed to bridge the gap between the history of science and the history of technology, this series publishes the best new work by promising and accomplished authors in both areas. In particular, it offers historical perspectives on issues of current and ongoing concern, provides international and global perspectives on scientific issues, and encourages productive communication between historians and practicing scientists.

W. Henry Lambright

NASA and the Politics of Climate Research

Satellites and Rising Seas

W. Henry Lambright
Maxwell School
Syracuse University
Syracuse, NY, USA

ISSN 2730-972X ISSN 2730-9738 (electronic)
Palgrave Studies in the History of Science and Technology
ISBN 978-3-031-40362-0 ISBN 978-3-031-40363-7 (eBook)
https://doi.org/10.1007/978-3-031-40363-7

This Palgrave Macmillan imprint is published by the registered company Springer Nature Switzerland AG.
The registered company address is: Gewerbestrasse 11, 6330 Cham, Switzerland

Paper in this product is recyclable.

To Nancy—who helped make this book possible.

Preface

Over the course of my career, I had written a book about the Moon (*Powering Apollo: James E Webb of NASA*) and Mars (*Why Mars: NASA and the Politics of Space Exploration*). I wanted to write about the home planet, Earth. The question was how to link it with my space interests. I felt climate change was the most important potential link, but climate change was so diffuse, and so much was already written by others, that it did not appear to be that great an option. Then I began exploring various dimensions of climate change and hit on sea-level rise. The oceans were so vast that they could not be fully observed, much less scientifically investigated, without space satellites.

Further reconnaissance revealed that NASA had been studying sea-level rise from space since 1992 beginning with TOPEX/Poseidon. Then came a series of follow-on satellites, thereby producing an uninterrupted multi-decadal record. Thanks in large part to satellites, NASA and scientists could measure how much the ocean had risen and how fast it was rising. Satellites could also help determine the factors causing rise, especially the melting of ice at the poles. There was a clear link between space policy and climate change. The clarity was critical, and it followed that a book on the NASA program was possible.

Next, I had to get resources to research and write this book. A small grant from Syracuse University in 2019 got me started on research and helped me prepare a formal proposal to write a book about NASA's role in climate change and sea-level rise research. In 2020 NASA awarded me a grant to prepare a history of this subject and gave me complete freedom to do this work. Thanks to the covid epidemic, there were a series of obstacles to achieving this task, especially putting a limit on travel. But I was able to persevere with numerous interviews via Zoom and telephone. I also had the help of various student-assistants along the way: Charlotte Hallett, Duncan Wood, Zack Bowman, Sophie Hernandez, Jillian Carafa, Miranda Nicole Nemeth, and Reilly Zink. Tammy Salisbury, an administrative assistant, also provided technical aid and a pleasant office atmosphere.

Along the way, I had the assistance of many others, such as my NASA Program Officer, Nadya Vinogradova Shiffer, NASA Historian Bill Barry, and various archivists at the NASA History Office. I also wish to thank the many individuals inside and outside NASA who gave their time for interviews. There were too many to list separately.

I am grateful also for the assistance of the Palgrave Macmillan editors and their associates, including Roger Launius, Jim Fleming, and Philip Getz.

My sons, Dan and Nat, Dan's wife, Sue, Nat's girlfriend, Stephanie, and my grandchildren, Ben, Katie, Bryce, and Darius, have been a source of inspiration. I owe a special sense of gratitude to my wife, Nancy. She took many burdens from me, kept me going in spite of setbacks, including some due to the covid pandemic, and provided support in a thousand and one personal ways. To all who helped, directly or indirectly, I hope the final product is worth your efforts. Any errors are my responsibility.

Syracuse, NY, USA W. Henry Lambright

Contents

CHAPTER 1

Introduction

Abstract This chapter details the monstrous challenge of sea-level rise as an impact of climate change. It discusses the network of satellites that the National Aeronautics and Space Administration (NASA) has developed and deployed over the years to determine what is happening, how fast, and why. It asks how this network was constructed and the role of bureaucracy-driven decision making in a difficult political environment.

Today, there exists an integrated, large-scale satellite system to track sea-level rise, its spread, causes, and impacts. In many ways, sea-level rise is the clearest and most understandable result of a warming planet. NASA and its partners built the network satellite-by-satellite over decades. It has gone from initiation to institutionalization. The introduction tracks how this network was built and the role of NASA in that "political construction." It shows how NASA drove decision-making—sometimes well, sometimes not so well. It demonstrates the barriers along the way and how NASA and its key leaders overcame them through various strategies to fashion a coalition of support and neutralize opposition.

Keywords Sea-level rise • Climate change • Satellites • Political environment • Key leaders • National Aeronautics and Space Administration (NASA)

W. H. Lambright, *NASA and the Politics of Climate Research*, Palgrave Studies in the History of Science and Technology, https://doi.org/10.1007/978-3-031-40363-7_1

On November 21, 2020, NASA launched a satellite named Sentinel-6 Michael Freilich from Vandenberg Air Force Base in California. The launch was a culmination of a multi-decadal program to monitor rising seas due to climate change. It was also a turning point. Like predecessors, the satellite bounced pulses of radar off the ocean surface to determine height.[1] What was new, aside from technical improvements, was the fact that NASA's chief partner in this endeavor was the European Space Agency (ESA) and that a twin of the satellite was already authorized for later in the decade. For the first time, advocates of satellite-based ocean observation could rest easy. There was a sense that after years during which every launch could be a last, a long-term process of technical and governmental innovation seemed to have reached a point of institutionalization. What was new was becoming routine, accepted as an ongoing governmental responsibility—even an international duty—to surveil the ocean and coastal waters to determine impacts of this aspect of climate change and help devise mitigation strategies.

Also, NASA was simultaneously collaborating with France, and to a lesser extent Canada and England, on the Surface Water and Ocean Tomography (SWOT) satellite, the first of a next generation of sea-level satellites, that would launch in December 2022. Moreover, NASA was in addition tracking how the melting of glaciers and other massive ice-masses at the poles were accelerating sea-level rise. Quietly, often guardedly, and with little fanfare, NASA, a space agency, had evolved a mission to Earth, and particularly climate change, with sea-level rise its bellwether. It had been a struggle to do so, not just technically and organizationally but politically. How did it happen?

The Sea-Level Threat

From 1992, when the first satellite capable of measuring changes in ocean height launched until 2022, global mean sea-level has risen by 10.1 centimeters (3.98 inches) around the world.[2] A recent international study has warned that by 2100 sea level could rise as much as 1.1 meters (or 3.6 ft).[3] Some observers believe that high-end view is too optimistic and predict six feet possible by then.[4] The U.S. government has forecast that coastal sea levels in this country will rise by one foot or more on average by 2050. If that happened, it would mean that an increase that took 100 years in the twentieth century would take about 30 in this century.[5] Satellite and other data have made it possible to predict what will likely happen, at least in the relative near term, but, even so, there are many unknowns. These include the fact that land is subsiding in some locations, adding to the concerns of destructive impacts on the coasts.

Thermal warming and expansion of the oceans accounts for part of the rise. The most important causal unknown has to do with what is happening at Earth's poles, where global warming is having its most dramatic impacts. Glaciers are melting at an alarming rate. If extreme case scenarios take place, the predictions about sea-level rise will change abruptly for the worse.

The prediction of an average foot rise by 2050 may not seem that much to the typical citizen. But what is average may be very different in particular places. Coastal cities such as Miami, New York, and New Orleans are at great risk as they already experience flooding and other impacts from storm surges and severe storms. Millions of people and trillions of dollars are at stake. While climate change amelioration through emission controls is essential, the United States and world have moved gradually into an era of adaptation to climate change, with sea-level rise alone likely to displace millions of climate refugees as the impacts of global warming come closer to where people live.

The oceans cover 70% of the Earth's surface. What will happen to American coasts will occur also in countries around the world, and probably inundate low-lying island states. This is an urgent problem, not going away, and is still being denied or ignored by powerful individuals in the United States and elsewhere. Denial has politicized the science of climate change. It has made NASA's mission to Earth by far its most controversial program. There are those inside and outside the agency who wish NASA concentrated only on planets like Mars and beyond, not the home planet. They ask: Why should a "space agency" have a mission for sea-level rise?

Satellite Observation

NASA did not intend to deal with sea-level rise when it first trained its satellites on the oceans with Seasat in 1978. While climate change and sea-level rise were postulated by some scientists and government officials, NASA was simply trying to determine if satellites could do anything useful involving the ocean. The agency had pioneered satellites for weather and land, so it was natural, said advocates, to extend this capability to the oceans. Most oceanographers were satisfied with existing research tools, such as ships, buoys, and tide gauges. Many said NASA was simply looking for missions to survive in the post-Apollo era.

It was not until 1992 that NASA launched a successor to Seasat, TOPEX/Poseidon. It did so in alliance with the French space agency, Centre National d'Etudes Spatiales (CNES). The precision of the measurements astounded NASA and many scientists. All of a sudden, they realized the world had a revolutionary technical capability. It was possible to observe the ocean and its dynamics as never before. Climate change was slowly emerging on the nation's agenda. NASA and a number of scientists decided the agency had to continue to gather data, because climate changed slowly and the seas along with it. They needed years of observations to see trends. Existing atmospheric satellites showed oceans getting warmer, and scientists knew that heat could make water expand. TOPEX/Poseidon was able to measure how much was the expansion. NASA instituted a sea-level rise effort as a leading edge of a broader "Mission to Planet Earth (MTPE)," as it was called before politics made NASA change the name to the more prosaic "Earth Sciences."

Along with its allies, chiefly France, NASA set about a long-term program of sea-level satellites—the Jasons—that led eventually to Sentinel-6 Michael Freilich, with the multi-national European Space Agency. There was no guarantee that this program would continue beyond TOPEX/Poseidon when it started. How it did so is a story of bureaucratic persistence aimed at an uninterrupted flow of vital information. The total program concerned with sea-level rise came to include satellites to investigate causes, particularly the melting of huge ice sheets and glaciers in Greenland and Antarctica. The key satellites—ICESAT, Grace, and their successors—became part of a larger technological system for understanding climate change.[6]

NASA was always ambivalent about a mission that seemed to require "continuity" in observations over decades. NASA called this activity "operations." It seemed to constitute routine monitoring. What NASA did was research and development (R&D). NASA scientists and engineers, particularly in centers like its Jet Propulsion Laboratory (JPL), were accustomed to working on the next technology, so as to answer novel science questions, and handing off what they developed to user agencies, such as the National Oceanic and Atmospheric Administration (NOAA). NOAA wanted this "other half" of the satellite mission, operations. Its problem had to do with money, and NOAA's lack thereof, along with demands from its vital core mission of weather prediction and inconvenient location in the Department of Commerce.

Sea-level rise was thus a case in which technology and societal need owing to climate change outpaced governmental missions, roles, attitudes, and funding. The quest of NASA to determine what it should do about sea-level rise and what other agencies should do, whether in the United States or abroad, is part of the history that this book traces. The nation has been egregiously slow to restructure government for climate change and to develop, use, and finance a technological system to observe continuously this existential problem, much less guide solutions. What is clear is that NASA has played a lead research role in dealing with this particular component—sea-level rise—but it has often been uncomfortable with the extent of this role and its politics.

Bureaucratic Entrepreneurship

This is not a study of the science and technology of sea-level satellites, per se. It is a book about government and the dynamics of institutional innovation—the making of a science and technology mission. It entails decision-making, primarily at the bureaucratic level of American government. If one contemplates "levels" of policy, one can see a macro-level in which the President and Congress make "national policy." One can also imagine policymaking at a micro-level inside a particular agency. At the mid- or meso-level is administrative (i.e., bureaucratic) policymaking—an agency working to do something new in a political environment involving the White House, congressional committees, the scientific community, various domestic agencies, international actors, and many others.

This is a dynamic subsystem concerned day-to-day with a sector of policy. Bureaucratic policymaking entails actions related to the micro-level and the macro-level of government, but its focus is best defined by the agency, those that lead it, and its immediate and most proximate political environment.[7] This is where NASA satellite policy has its locus. It is the level where most of the satellite policy decisions have been made, most of the time, since NASA became a moving force for satellite innovation.

Policy change and its dynamics can be studied in various ways. But the most relevant to this effort is a process model.[8] In this approach, there are stages. First is *agenda-setting*. A problem of opportunity gets on the agenda of a government agency. Second, officials in the agency, often with others outside the agency, engage in *formulation* activity determining options for policy actions.

It is not enough for agencies to act alone to get to the *adoption* stage of policy. They need the legitimacy and resources that come from accountabilities to the macro-political level of the President and Congress. If authoritative actors agree, the process moves to *implementation*. The agency executes. Along the way, there can be an *evaluation* stage that can lead to a reorientation of agency tactics. Finally, this model assumes the implementation continues to the *institutionalization* stage, where what was new becomes routine. It is applicable to any individual satellite project and to the program as a whole. This study deals with individual projects, but in the context of a larger program stretching over decades.

NASA's clearest role is that of a research and development agency, a role that Congress bestowed on it in the Space Act of 1958. It required that it be a force for technological innovation in the realm of space and aeronautics. The nation was embarrassingly behind its Cold War rival. Congress wanted it to catch up and surpass the Soviet Union. How it applies that role to particular missions is a matter of interaction between NASA and its political environment. That is as true for its dominant activity, namely human space flight, as it is for its programs of robotic spacecraft and space telescopes. The culture of NASA favors accomplishing "firsts." Obviously, it does more than that, but the agency and those who set its tone want most to advance in science and technology.

To move a new technology from stage to stage in the policy process, NASA has to be more than an agency carrying out an assigned task. It has to be a political actor, working to get resources and other decisions. As a political actor, NASA has to build an "advocacy coalition" sufficient to move technology forward. Otherwise, *termination* interrupts the innovation process and ends it.

Ideally, all the actors necessary to reach institutionalization would be aggregated at the agenda-setting stage, and they would agree to a course of action and stay united through the entire process. Reality is not so neat. The course of policy is seldom as straightforward as the process model suggests. There are gaps, zigs, and zags along the way. It usually will have opposition of one kind or another. In many cases, it seems a miracle when a new science and technology program reaches fruition and becomes routine. Bureaucratic entrepreneurship, like private entrepreneurship, entails risks. The risks are technological and financial. They are also political.

This has emphatically been the case with NASA's mission entailing climate change. There were participants in the policy process who were nonbelievers in climate change, or at least change due to human actions. They saw research about it, such as sea-level rise, as a waste of money or threat to their interests. They opposed NASA speaking out and playing an active research role in this field. There were others with the opposite view. They wanted NASA to move more aggressively toward application, mitigation, and "solutions." This dynamic has been gradually changing in favor of more acceptance of climate change and its impacts along with a duty for outreach. But during most of the period studied here, the subject was intensely conflicted and politicized.

NASA over years carved its role amidst these countervailing pressures. Leaders of the earth science/sea-level rise mission have had to make choices. Writers about science policy make a distinction between what they call "policy for science" and "science in policy."[9] For NASA, that comes down to choice about to what extent it should concentrate on building a science and technology program aimed at producing new knowledge, "a predictive capability," rather than pushing that knowledge deeply into policy solutions. The latter is an additional part of the mission. It applies the satellite technology and also the science derived from it to tackle explicit policy remedies, including regulation.

The choice of NASA leaders generally has been to emphasize "policy for science" and limit "sounding an alarm" to trigger climate change policy actions. One NASA scientist in particular, James Hansen, sought to sound an alarm to galvanize politicians, the media, and public.[10] The problem was that many in NASA believed this kind of science in policy strategy endangered the science research program politically, given the influence of climate change deniers in Congress. NASA aimed to communicate the risks it saw, but less stridently, often letting the science speak for itself. This is thus a story of what NASA did in managing its sea-level program: how, why, and why not in a difficult environment.

What NASA leaders of this program, or any program, want is autonomy, leeway to make decisions.[11] The most they can get is "semi-autonomy," enough leeway to shape a program, subject to multiple and often conflicting pressures. These accountabilities come from the president, Congress, the scientific community, interest groups, the media, administrative partners in the United States and abroad, and the general public. Establishing an innovative and controversial program in government requires sustained bureaucratic entrepreneurship. The leadership begins with individuals inside the agency often in league with likeminded outside forces who win the support of agency superiors and the organization generally. They make their interest, agency interest. The agency then seeks outside support in national policy.

Bureaucratic leaders hone their expertise, build internal and external constituencies, find collaborators, construct coalitions of support, and seek to neutralize opponents. They do all this to move their technological programs forward. "Failure is not an option" goes the NASA refrain. But pure success does not always happen either. Compromise is the rule as bureaucratic entrepreneurs seek institutionalization of their programs. That doesn't end political conflict, but it is a conflict that is then largely contained by a strong standing constituency that secures the program through a coalition of commitment.

To reach this point takes much time and requires a relay of administrative leaders and successive advocacy coalitions. Each leader moves the process ahead incrementally, stage by stage, satellite by satellite, to achieve the program goal of institutionalization. Some do better than others, and all muddle through at times. This takes years of persistence as presidents and congresses come and go and adaptation of means to attain a chosen end. The result after more than 30 years of data is better understanding the impact of humans on the Earth's climate, especially its seas. "The rise of sea level caused by human interference with the climate now dwarfs the natural cycles," stated Josh Willis of NASA's Jet Propulsion Laboratory, an oceanographer and project scientist for Sentinel-6/Michael Freilich. He adds: sea-level rise "is happening faster and faster every decade."[12]

The Narrative Ahead

This book, therefore, is a political/policy history of a NASA program to observe, understand, and in some ways advocate and help mitigate the vexing challenge of climate change as it relates to sea-level rise. Sentinel-6 Michael Freilich climaxed but did not end the program. It marked the end of its beginning. SWOT starts another chapter, one even more relevant to coastal threats, as will successors to its polar ice satellites. NASA and its allies in the United States and abroad have created not only a technology but an integrated infrastructure for a vital climate mission. How that happened illuminates the way government, science, international relations, and especially bureaucratic leadership interact in a charged political environment.

A relay of administrative leaders largely unknown to the public has forged a major climate change/sea-level rise role for NASA over several decades. That role involves pioneering research and satellite technology development. It entails efforts to communicate science to the policy community and public. It embraces struggles over how far to go beyond research to long-term monitoring, what NASA called "operations."

What NASA has done has been a function of bureaucratic maneuvering in a context of intermittent opportunity and frequent political risk—i.e., entrepreneurial leadership. The result after many years is a capability to understand, predict, and help alleviate the impacts of sea-level rise across the globe and regionally. How did NASA build this capability? Who did what? Administrative leaders at NASA, mostly high-ranking bureaucrats, connected science, technology, and politics to make progress in policy. They sought to collaborate with agencies in the United States and abroad to move a multi-decade program forward. What they did took place in an environment of political conflict and inchoate national policy. In the end they forged an institutional commitment spanning the Atlantic and built a program of mission-driven research.

Notes

1. Paul Voosen, "Seas are Rising Faster Than Ever," *Science* (Nov. 20, 2020), 901.
2. "Tracking 30 years of Sea Level Rise," *NASA Earth Observatory*. Retrieved from https://earthobservatory.nasa.gov/images/150192/tracking-30-years-of-sea-level-rise
3. "New high-end estimate of sea-level rise projections in 2100 and 2300," *WCRP*, Oct. 25, 2022. Retrieved from https://www.wcrp-climate.org/news/science-highlights/1955-new-sea-level-projections-2022#:~:text=The%20high%2Dend%20global%20mean,for%20strong%20warming%20in%202100
4. Jeff Goodell, *The Water Will Come: Rising Seas, Sinking Cities, and the Remaking of the Civilized World*, (N.Y.: Little, Brown, and Co., 2017).
5. Henry Fountain, "Coastal Sea Levels in U.S. to Rise a Foot by 2050, Study Confirms" *The New York Times*, (Feb. 15, 2022), A19.
6. Paul Voosen, "NASA Set to Announce Earth System Observatory," *Science*, (May 7, 2021), 554.
7. James Q. Wilson, *Bureaucracy: What Government Agencies Do and Why They Do It* (New York City, NY: Basic Books, 1989).

8. Paul Sabatier Ed., *Theories of the Policy Process*, (Boulder, CO: Westview 2017).
9. Homer Neal, Tobin Smith, and Jennifer McCormick, *Beyond Sputnik: U.S. Science Policy in the 21*st *Century* (Ann Arbor: University of Michigan, 2008), 10–11.
10. James Hansen, *Storms of My Grandchildren: The Truth about the Coming Climate Catastrophe and Our Last Chance to Save Humanity* (NY: Bloomsbury, 2009).
11. Daniel Carpenter, *The Forging of Bureaucratic Autonomy: Reputation, Networks, and Policy Innovation in Executive Agencies, 1862-1928*, (Princeton, NJ: Princeton University Press, 2001).
12. "Tracking 30 years of Sea Level Rise," *NASA Earth Observatory*. Retrieved from https://earthobservatory.nasa.gov/images/150192/tracking-30-years-of-sea-level-rise

CHAPTER 2

The Coming of Seasat

Abstract The origin of NASA's interest in the oceans is tracked from the agency's early years (1960s) to the launch of the first significant ocean satellite, Seasat, in 1978. This chapter shows NASA Administrator James Fletcher's role in the 1970s and later (1980s), in pointing NASA toward environmental missions. It discusses NASA's connection with the National Oceanic and Atmospheric Administration (NOAA) and weather satellites and difficulties of the "hand-off" of satellites from development to operations when two or more agencies are involved. Seasat lasted only three months before failing but showed the high value of ocean satellites to various potential users, particularly scientists.

Keywords Seasat • Ocean satellite • National Oceanic and Atmospheric Administration (NOAA) • James Fletcher

The ocean satellite program grew slowly, tortuously, out of NASA's early efforts to apply space technology to Earth. From the advent of the agency in 1958, NASA's remote-sensing engineers yearned to extend space to all places possible. Everywhere they looked, they saw opportunity as technological revolutionaries. NASA's top leaders, including James Webb, Administrator in the 1960s, believed a "program" could help NASA with political support. That was one reason NASA organized an Office of Space

W. H. Lambright, *NASA and the Politics of Climate Research*, Palgrave Studies in the History of Science and Technology, https://doi.org/10.1007/978-3-031-40363-7_2

Science and Applications (OSSA) to facilitate technology transfer from space to Earth. NASA believed that once it demonstrated the wonders of space technology, users would flock to its "spin-offs." This view proved naïve.

The Apollo Illusion

Part of NASA's illusion derived from Apollo's success. The Apollo model was one in which NASA developed, operated, and used its own hardware. This coupling of roles, backed by a national mandate and policy consensus, allowed NASA to set its own technological and operational objectives and worry little about other agencies. With strong leadership and ample autonomy, the agency achieved the seeming impossible with a Moon landing in 1969.

But missions involving transfer to others needed a very different model, even where the perspective of space clearly offered advantages over conventional technologies. NASA was not necessarily the operator or ultimate user of the new systems it developed. It was, therefore, not the final judge of whether what was developed was acceptable. It would not be enough to establish technical feasibilities. What was needed was to match what NASA did with what others—operators and users—wanted, both technically and financially.

The issue of transfer was compounded by what was often an inequality of organizational capacity and political power between the agency providing the technological push and that expected to pull for use. Operational agencies had their own missions, funding streams, and interests. The innovation process was about more than research and development—NASA's strength and "distinctive competence." It extended to the incorporation of a new technology into the routines of the operator and user. Often, the relations among developer, operator, and user were complex because they were different entities and what if there was no existing user particularly interested in space?

Initiating

In the 1960s, NASA pushed to open the new frontier of space technology wherever it could. It was quite successful, early, with communication satellites. Weather proved more complicated, land even more so, and the oceans were especially difficult.

In the case of weather, NASA launched a satellite called TIROS in 1960. For the first time, meteorologists saw large weather patterns from space on a scale previously impossible. In 1961, a TIROS satellite helped track a dangerous hurricane, Carla, as it approached the Gulf Coast. The early warning led to the evacuation of more than 330,000 people and saved countless lives. The would-be operator and user, the Weather Bureau, realized it had a potentially revolutionary new tool.

NASA and the Weather Bureau, located in the Commerce Department, agreed on a partnership such that NASA would develop new satellites and the Weather Bureau would take control and fund them once they became operational. In the mid-1960s, NASA moved ahead with the next-generation satellite, NIMBUS. The Weather Bureau refused to accept NIMBUS, declaring it was too expensive. The Department of Defense (DOD), another possible operator/user, backed the Weather Bureau. The White House arbitrated. The compromise was that NASA would upgrade TIROS for the Weather Bureau and then transfer the operational machine to the Weather Bureau, which would pay for it. At the same time, NIMBUS would continue, under NASA, for research users, who wanted a more sophisticated satellite. DOD looked to its own devices.

Where the weather satellite was concerned, there was an operational pull for a less expensive satellite than NASA preferred. There was bureaucratic struggle, but the agencies found a way to cooperate. When it came to remote sensing for land resources which NASA also developed in the 1960s, transfer was even more problematic. There were two agencies that showed interest: the Department of Interior and the Department of Agriculture. In 1966, Interior was so enthusiastic that it complained that if NASA did not move more quickly with developing this land satellite, it would create one itself. The warning was a ploy, but it did get NASA to accelerate development.[1]

Thinking About the Oceans

Weather and land were among the initial Earth applications. NASA also thought about the oceans, NASA had few if any oceanographers in its employ when it began—a problem. There was also no operational "ocean agency" asserting its claims. Ocean interests in the government and private sector were quiet. The Navy had other priorities. NASA found it difficult to get the attention of the scientific community when it came to the oceans. In 1964, NASA sponsored a conference at the Woods Hole

Oceanographic Institute in Massachusetts. One hundred fifty oceanographers attended. Gifford Ewing, Director of Woods Hole, chaired the meeting. In a 1965 report, Ewing called the meeting's subject "oceanography from a satellite" as having an "incongruous" sound to ocean scientists. But he wrote it was "unthinkable that oceanographers will not find ways to exploit this burgeoning technology for the advancement of their science." He called the oceanographers' attitude at the meeting one of "restrained enthusiasm." They thought the most feasible application would be tracking buoys for navigational progress.[2]

In 1969, NASA sponsored another oceanography conference, this time at Williamstown, Massachusetts. It gave particular attention to radar altimetry, an emerging technology to measure distances from a satellite to the ocean surface. Some thought it might help in studying large ocean currents, such as the Gulf Current, a major priority in ocean science. The Williamstown conference concluded that if NASA could get altimetry accuracy down to 10 cm (4 inches), it could be possible to map the ocean's "dynamic topography"—surface features. But observations would have to be not only incredibly precise; they would also have to be implemented over a very long time.[3] Change in the ocean could take decades to be measured and proved. These two conferences and others provided requirements for NASA's engineers to reach if they wanted to apply space technology to the ocean in a way scientists would find useful.

Fletcher's Vision

In 1971, James Fletcher became NASA Administrator. The next year he established the Earth and Ocean Physics Application Program under OSSA. The action was part of an aggressive move by Fletcher into Earth Science and environmental fields. A physicist and former university president, Fletcher wanted to keep NASA viable as a major agency in the 1970s as it struggled after Apollo. He secured a decision to build a space shuttle from President Richard Nixon in 1972 that continued human space flight. However, he also made it clear that he wanted NASA to be an "environmental agency." Toward that end, NASA launched an Earth resources satellite, Landsat, in 1972.

The potential user-agencies, Agriculture and Interior, could not get funds from Nixon to take it on, however. Some policymakers in the White House and Congress thought it should be operated by the private sector. The consequence was that Landsat became a long-term orphan in search of an institutional home. It eventually wound up with NASA in partnership with Interior's U.S. Geological Survey. They co-managed and co-funded the enterprise. That was long after Fletcher had presided over its launch.

While frustrated with Landsat, Fletcher was serious about applying space to Earth, including the ocean. He publicly referred to NASA as an "environmental agency." Fletcher told Congress in 1973 that while "NASA is called the Space Agency…we could be called an Environmental Agency…everything we do helps in some practical way to improve the environment of our planet and helps understand the Earth and its environment."[4] Historian Roger Launius attributed Fletcher's environmental activism to his Mormon sense of stewardship. There may also have been bureaucratic needs involved as well. The 1970s were a time when environmental values loomed large—the Environmental Protection Agency (EPA) was born in 1970—and NASA needed to show its relevance to national interests.

Ozone Depletion

In 1975, Fletcher gained a congressional warrant to establish a new environmental research mission to study ozone depletion. This was a technical controversy at that time, and NASA argued it had the satellites, as well as overall competence, to best study a global problem.

Significantly, NASA won this mission in a contest with a new agency, the National Oceanic and Atmospheric Administration (NOAA), established by Nixon in 1970. NOAA, with the Weather Bureau a core unit, was born following a long effort by marine interests in having a civilian ocean agency. However, these interests could not wrest a large ocean-oriented entity, such as the Coast Guard, as a NOAA unit. Nor could they get ample funds for research and development. Heralded by proponents in the late 1960s as a potential "wet NASA," NOAA could not get extracted from the Commerce Department and be given independent agency status, like NASA, with an administrator reporting directly to the President. It was born with ambition, but a relatively small budget. Inheriting the Weather Bureau, meteorology was far more powerful within NOAA than oceanography.

NASA, with a reputation borne of Apollo and a highly supportive congressional constituency, won the mission of ozone depletion, and, with it, legitimacy to expand its role as an "environmental agency." In June 1975

Congress passed legislation directing NASA "to conduct a comprehensive program of research, technology, and monitoring of the phenomena of the upper atmosphere." This was a significant, all-embracing research mandate. It was far more than engineering. The legislation also enlarged NASA's role from R&D to "monitoring," thereby making it an "operator." As such, it could also be the prime communicator of science in this field to policy makers. In this discrete area, ozone depletion, NASA was the dominant science and technology agency.[5] Ozone depletion was also "policy-relevant" science. Politicians expected NASA to provide useful information.

Under Fletcher, the small band of ocean advocates in OSSA had clear sailing to think about ocean observation as a new mission for NASA as an "environmental agency," NOAA's ambitions notwithstanding.

Stepping Stones to Seasat

The Williamstown conference had given NASA an altimetry target toward which to strive for accuracy—10 cm. NASA moved toward that mark incrementally, while testing different kinds of sensors for color, wind, and other ocean features.[6]

First came the launch of Skylab in 1973. This was a small space station composed of Apollo components aimed to show that astronauts could do useful work in Earth's orbit. Ocean sensors, including altimetry, were part of the exercise. Unfortunately, damage at the time of launch limited what various instruments could do. However, the altimeter did show some promise.

NASA next planned a small satellite, GEOS-3, for a 1975 launch. This would focus around altimetry to see what it could do with an enhanced accuracy NASA believed it could attain. Then would come Seasat-A in 1978. Seasat-A would mark a further advance in precision and reach the goal the scientists wanted. It would be a proof of concept mission in NASA parlance. Seasat-A would be NASA's first satellite dedicated specifically to a wide range of possible oceanography uses—wind, currents, tides, and more. NASA hoped Seasat-A would have a follow-on, Seasat-B. With the weather satellite in mind, advocates expected users to pay for operational satellites, whether Seasat-B or later versions. They saw NOAA and DOD as possible users.

NASA intended scientists to be users as well. It established an advisory body of scientific researchers to help it. It also sought advice (and endorsement) from the most prestigious scientific body, the National Academy of Sciences. One scientist it hoped to interest was Carl Wunsch.

Enlisting Carl Wunsch: A Scientific Gatekeeper

Wunsch was a leading oceanographer at MIT. Well-connected scientifically and in science policy circles, Wunsch had not participated in NASA's 1960s conferences or other NASA advisory groups relating to the oceans. He in fact knew little about satellite remote sensing and what he knew did not impress him. Nor did the arguments of ocean scientists like Ewing, who were early proponents of the technology. Wunsch was exactly the kind of skeptical and high-profile oceanographer NASA needed on its side to attract a scientific constituency.[7]

In 1974, Wunsch received a call from a representative of the National Research Council (NRC), the working arm of the National Academy of Sciences, asking him to join its Seasat panel. Wunsch responded with an empathetic "No!" He stated that NASA's efforts were all hype. "I thought NASA's public relations machinery was far outstepping the importance of its contribution," he recalled. NASA, in his view, was a solution looking for a problem. Wunsch eventually agreed to serve, but he expected to show how little use NASA's remote sensing would be to scientists like himself.

Once he perused some of the material NASA supplied, however, especially about altimetry, he "got interested." Space-based altimetry might indeed be a real solution to genuine problems oceanographers encountered. Indeed, he thought oceanography faced a crisis.

One aspect of the crisis was that recent research showed that the oceans were much more turbulent than oceanographers had believed to have been the case. Somehow, they had to investigate that turbulence and the currents at the surface and below. Another was the emerging issue of climate change. A senior oceanographer, Roger Revelle, was urging his colleagues to take it very seriously. Revelle had presciently declared in 1957: "Human beings are now carrying out a large-scale geophysical experiment of a kind that could not have happened in the past nor be repeated in the future."[8]

As Director of the Scripps Institution of Oceanography in California, Revelle had supported the work of Charles Keeling who had measured the steady accumulation of CO2 in the atmosphere. The "Keeling Curve" would ultimately become an iconic marker of climate change. Revelle wanted oceanographers to study the ocean's role in climate change and alerted them to possible dangers of climate change ahead. Yet another worry Wunsch had about his field was competition from meteorologists who were carrying out large-scale atmospheric experiments and getting into atmosphere–ocean interactions without much attention to oceanography.

The chief challenge, Wunsch believed, was "observational." Getting at the issues of large-scale turbulence and climate change required a much more synoptic approach than the ships and other surface tools oceanographers used. He knew that meteorologists disparaged oceanographers for depending on "swamp models." Their view was that oceans didn't change much over long periods of time and were not important, at least relative to atmosphere, in climate. Wunsch thought the meteorologists were wrong, but how to prove that fact was an issue.

He believed oceanographers had to think bigger with tools worthy of the challenge, and he felt satellite altimetry just might be that tool. The oceanographers' dominant technology of choice—ships—could not "see" the large and powerful currents whose understanding was central both to ocean science and climate. If satellite altimetry could get sufficiently precise measurements of sea surface height, in his view, scientists could combine them with other measurements of the Earth to produce unprecedented knowledge of the seas and their impacts. Wunsch decided to learn more about what NASA was proposing with Seasat.

Seasat's Problems

While seeking advice (and interest) from ocean scientists like Wunsch, NASA moved ahead with its next step in ocean satellite development: the launch of GEOS-3—in July 1975. GEOS-3 was all about altimetry and whether NASA's effort to improve accuracy was succeeding. The answer was: Yes. GEOS-3 was able to measure significant ocean features with 20 cm (8 inches) precision. NASA wanted to get down to the 10 cm (4 inches) mark with Seasat-A in 1978.

But, in 1977, Congress's financial watchdog, the General Accounting Office (GAO), reported to Congress that NASA had over-promised in getting Congressional approval for Seasat-A. NASA had claimed operational agencies—potential users—would pick up the expense of ocean satellites after Seasat-A and Seasat-B, said GAO. GAO found no evidence that would happen. It advised Congress to end the Seasat program with Seasat-A.[9]

By this time, Fletcher was gone, but NASA had another ocean advocate at its helm. The new president, Jimmy Carter, had appointed Robert Frosch, a physicist and oceanographer, to head NASA in 1977. Frosch had been a top research director with the Navy and had served at the Woods Hole Oceanographic Institute. Frosch in turn appointed William Nierenberg, the successor to Revelle at Scripps, as Chairman of the NASA Advisory Committee (NAC).

Nierenberg, through NAC, advised Frosch that the oceanography community was not prepared for the torrents of data Seasat would likely produce. He advised Frosch to make a major effort to get ocean scientists ready for Seasat-A. He and NAC also recommended NASA centralize and consolidate its existing ocean program, which was decentralized and subcritical in size. The advice was to build an ocean science mission at two NASA centers, one in the West and one in the East.[10]

Seasat Fails

On June 26, 1978, Seasat-A launched. It carried five major sensors to measure winds, waves, currents, temperatures, ice, and atmospheric and water vapor. It was the world's first true comprehensive ocean satellite. Altimetry was one of the instruments. It showed the Gulf Current as never seen before and achieved its precision goal with 100% accuracy. After a life of 99 days, the satellite failed. A massive electrical short rendered it useless. Quickly, a conspiracy theory erupted—namely that the Department of Defense purposefully destroyed the satellite lest the Soviet Union discover how good the technology was for observing nuclear-carrying submarines.

The conspiracy theory lived in spite of NASA denials. What did not live was NASA's idea of Seasat-B and a potential operational program to follow by another agency. What was NASA to do? With Frosch, an oceanographer, in charge of NASA, it was obvious that NASA's move into the ocean environment was not going to end with the death of Seasat.

Notes

1. Homer Newell, *Beyond the Atmosphere: Early Years of Space Science* (Washington, DC: USGPO, 1980), Ch 19.

2. G.C. Ewing, *Oceanography from Space.* Woods Hole Oceanographic Institution. Proceedings of a Conference held at Woods Hole, 24–26 Aug. 1964. Woods Hole, Mass. (1965).
3. Massachusetts Institute of Technology, *The Terrestrial Environment: Solid Earth and Ocean Physics, Application of Space and Astronomical Techniques* (Cambridge, MA: 1989). See also W. Stanley Wilson et al. "Satellite Oceanography—History and Introductory Concepts," in *Encyclopedia of Ocean Sciences*, 3rd Ed, vol. 5, (2019), 347–361. See also Erik Conway, "Drowning in Data: Satellite Oceanography and Information Overload in the Earth Sciences," *Historical Studies in the Physical and Biological Sciences*, vol. 37, No. 1 (Sept. 2016), 127–151.
4. Roger Launius, "A Western Mormon in Washington, D.C.: James C. Fletcher, NASA, and the Final Frontier" *Pacific Historical Review* (1995), 236.
5. W. Henry Lambright, *NASA and the Environment: The Case of Ozone Depletion* (Washington, DC: NASA, 2005), 7–8.
6. W. Stanley Wilson et al. "Satellite Oceanography—History and Introductory Concepts," in *Encyclopedia of Ocean Sciences*, 3rd Edition, vol. 5, (2019), 347–361; See also Erik Conway, "Drowning in Data: Satellite Oceanography and Information Overload in the Earth Sciences," *Historical Studies in the Physical and Biological Sciences*, vol. 37, No. 1 (Sept. 2016), 127–151.
7. Carl Wunsch, "Toward the World Ocean Circulation Experiment and A Bit of Aftermath," in M. Jochum and R. Murtugudde, Eds., *Physical Oceanography: Developments Since 1950* (NY: Springer, 2006), 181–201.
8. Spencer Weart, *The Discovery of Global Warming* (Cambridge, MA: Harvard University, 2003), 30.
9. Comptroller General of the U.S., Report to Congress. *The Seasat—A Project: Where it Stands Today* (Sept. 16, 1977). This office was renamed General Accountability Office in 2004.
10. Erik Conway, "Drowning in Data: Satellite Oceanography and Information Overload in the Earth Sciences," *Historical Studies in the Physical and Biological Sciences*, vol. 37, No. 1 (Sept. 2016), 127–151.

CHAPTER 3

Adopting TOPEX/Poseidon

Abstract NASA Administrator Robert Frosch, an oceanographer, saw the value of satellite observations and directed his agency to build a program for ocean satellites with research as a purpose. W. Stanley Wilson, an oceanographer and bureaucratic entrepreneur, joined NASA to create such a program (Seasat was seen as a technology demonstration only). This chapter details what and how Wilson built an ocean program. At NASA Administrator James Beggs' insistence, he acquired a research and development partner (France's space agency, *Centre national d'études spatiales*, CNES). They promoted a new satellite, TOPEX/Poseidon. One reason it took so long to go from Seasat to the launch of TOPEX/Poseidon in 1992 was rivalry between Wilson and Shelby Tilford, who headed a higher agency priority, the Earth Observation System (EOS). Tilford wanted the ocean and Wilson as part of EOS. Wilson wanted autonomy. Tilford won and Wilson departed before TOPEX/Poseidon launched.

Keywords Robert Frosch • W. Stanley Wilson • James Beggs • Centre national d'études spatiales (CNES) • TOPEX/Poseidon • Earth Observation System (EOS)

W. H. Lambright, *NASA and the Politics of Climate Research*, Palgrave Studies in the History of Science and Technology, https://doi.org/10.1007/978-3-031-40363-7_3

Although Seasat was a failure insofar as lasting only a little more than three months, it revealed as much about the surface of the ocean in its brief life as shipboard oceanography had produced in 100 years. It showed the Gulf Current and other large-scale oceanic movements as never seen before. While most oceanographers did not take particular note of Seasat, those who did were excited. Seasat achieved its goal of 10 cm accuracy. The oceanographer whose reaction counted the most was Frosch. He directed his NASA managers to "go make an oceanography program."[1] What existed was an application program, an engineering effort to demonstrate technology. Frosch wanted a more complete program, one entailing scientific research. That meant having an oceanographer in charge.

W. Stanley Wilson

The man NASA hired was W. Stanley Wilson, a scientist with a PhD in oceanography from Johns Hopkins, who had been running the Office of Naval Research's basic research program. At age 41, Wilson was experienced, energetic, and ambitious. One of his first actions following arrival in 1979 was to take a tour of NASA centers that had been involved in Seasat. Headquarters had been more a coordinator than manager, and Wilson intended to be a manager.

Wilson saw his task as merging oceanography and remote-sensing engineering. As a scientist, he wanted his principal associate at headquarters to be a remote-sensing engineer. Believing altimetry the most significant technology aboard Seasat, he made it a point to meet the man most responsible for altimetry on the satellite. This was Bill Townsend, who happened to be located at one of NASA's smallest centers, Wallops, on Virginia's eastern shore. Wilson was able to persuade Townsend to join him in Washington. Wilson called his program, located in OSSA, "Oceanography from Space."[2]

He now needed specific centers to provide technical depth to his program, centers that could perform research and help manage contracts. Prior to Seasat, Frosch's advisers had called for consolidating oceanographic R&D into two groups, one in the West and one in the East. Wilson took this guidance to heart. There were two major branches of oceanography: biological oceanography and physical oceanography. Wilson made the Goddard Space Flight Center in Maryland responsible for biological oceanography and put the Jet Propulsion Laboratory (JPL) located in California, in charge of physical oceanography, the latter being Wilson's specialty.

One of the reasons Wilson favored JPL was that he found Moustafa Chahine, who ran Earth Sciences at JPL, to be one who shared his desire to build a sustained ocean program. He also found him more likely to be responsive to his direction than officials at some of the other centers he visited. There was another reason he wanted to work with JPL. JPL had the lead role in NASA for planetary exploration. That meant it had substantial computer capacity for data management. As Frosch's advisers had noted, data management was going to be critical given how much data space satellites could produce, as shown by Seasat in its brief life. Very soon, Wilson and Chahine began hiring young ocean scientists to work on the new effort.

As he formed his team, Wilson reinforced actions already underway to assess the data Seasat had compiled. Seasat had millions of dollars set aside to operate for at least three years, but no money for research. This was because it was not a research program but a technology demonstration. Wilson took that money and redeployed it for analysis.

He sponsored a series of workshops to get the data analyzed and also help attract oceanographers to the new field of satellite-based oceanograpy. He made data available to all users, a shift in policy already underway when he arrived. Traditionally, NASA gave investigators directly involved in a project access first, while others waited. Also, significantly, some of the early users were in Europe. Some had even attended the Williamstown conference and had followed Seasat's development.

It was not just researchers who had paid attention to Seasat. This satellite was a trigger in awareness for operating agencies. NOAA and DOD proposed a National Ocean Satellite System (NOSS) to make ocean observations they wanted on a regular basis. They offered NASA 25% of the projected payload for research. NOSS would have been a $1 billion satellite, an interagency endeavor. It never got beyond planning and was killed in 1981.[3] Satellite oceanography remained a research and development endeavor under NASA and Wilson's auspices.

Connecting with Scientists

Whatever else he did, Wilson had to have a Seasat successor. He also had to have credible scientific advice about what this satellite needed to do. He reached out to Wunsch, who had moved from scientific skeptic to scientific champion of satellite oceanography, a view strengthened by Seasat. Wunsch became a key ally as Wilson constructed an advocacy coalition for his program. Wilson in early 1980 asked Wunsch to chair a Science Working Group (SWG) to determine scientific requirements for the next satellite mission. He wanted Wunsch to work closely with Robert Stewart, the first scientist JPL hired for ocean research. He made Stewart the project scientist for the Seasat successor.

Wunsch's interest was energized by recent participation on a National Academy of Sciences assessment of climate change, known as the Charney Committee, after its chairman, Jule Charney.[4] The panel estimated the future impacts of global warming given scientific knowledge at that time. The panel produced a grim prospect but found one of the largest gaps in knowledge was the role of the oceans in climate change.

Wunsch also saw opportunity to link NASA and satellites with a large international oceanographic research program being discussed at the time. One of the groups involved in planning was called the Committee for Climate Change and the Oceans (CCCO). Attending a meeting of this group, Wunsch made a "convincing scientific case for a global ocean circulation experiment."[5] Wunsch subsequently took a leadership role in what became the World Ocean Circulation Experiment (WOCE). WOCE needed a global technology, he said, and satellites could take a global view. With altimetry, satellites might help determine the ocean's impacts on climate change. Wunsch became the critical link between NASA and the oceanographic community.

Going Abroad

A number of Europeans were interested in satellites and the oceans. Late in 1980, Wilson attended an "Oceanography from Space" symposium in Venice. There he gave a "Vision" talk about his program. He declared: "Because ocean remote sensing is coming of age, our view of the next decade is an exciting one, with great potential for oceanography. Not only has the remote sensing community demonstrated a capability for observing the oceans from space, but they—together with the oceanography community—are well along the way in developing a capability to physically interpret the resulting observations."[6]

After his talk, a French scientist in the audience, Michael Lefebvre, came up to him. Expressing his fascination with the subject, he invited

Wilson to visit him at his lab at the Toulouse Space Center of CNES, France's space agency. Wilson accepted the invitation, went to Toulouse, and was briefed on French research in physical oceanography, remote sensing, altimetry, and related subjects. The two men related well, and Wilson decided to stay in touch with Lefebvre and CNES.

Wilson moved his new program forward in planning and constituency-building. It helped greatly that he had a mandate from NASA's administrator. But in November 1980, the election put a new president, Ronald Reagan, in office. That would mean a different NASA administrator. What would that portend for Wilson's nascent program?

Formulating a Program

As the transition from President Carter to President Reagan ensued, in early 1981, the Science Working Group, with Wunsch and Stewart in the lead, presented a report to Wilson regarding Seasat's successor, which Wilson and his team called "TOPEX." TOPEX stood for Topography Experiment. They stated that "The primary goal of the [TOPEX] experiment is to measure the topography of the ocean (the mean surface pressure) over the entire ocean basins for several years, to investigate these measurements with subsurface measurements and models of the ocean's density field, in order to determine the general circulation of the ocean and its variability." Moreover, it was to then "use this information to understand the nature of ocean dynamics."[7]

They called for observing ocean dynamics with 5 cm accuracy every 10 days for five years. They recommended working closely with WOCE. The aim was to have a flagship user in the loop as NASA developed the technology. Wunsch and Stewart declared that TOPEX had the potential to "revolutionize" ocean science, "solving problems that extended back centuries, that had eluded all attempts at solutions."

They also argued that TOPEX should not repeat Seasat's mistake. It should have money for scientific analysis and a data management facility with experts primed to translate huge amounts of data into scientifically usable knowledge. JPL already claimed the data management role although Wilson was thinking it might be better to give it to Scripps, where Nierenberg was in charge. Wilson was anxious to broaden the constituency for the ocean program's long-term survival.

Selling Science

It was increasingly clear to Wilson and others involved with TOPEX planning that they had a problem with the oceanography community. Wunsch found his oceanography colleagues on WOCE quite resistant to relying on satellite altimetry, which was foreign to most of them. “I’d much rather have another ship,” one prominent ocean scientist told Wunsch. “What we really need is a lot more floats [a surface-level instrument] near the gulf stream,” said another. Even Wunsch’s MIT colleague and mentor, Henry Stommel, was skeptical and told Wunsch’s wife her husband was “trying to destroy his career” by going in the direction he was heading. Wunsch visited Frank Press, Carter’s Science Advisor, before Press departed his position. Press warned him that the satellite oceanography program he contemplated was too small to win congressional support. Big programs got industry contractors involved, and they could lobby Congress. Lobbyists and politicians would not bother with smaller programs.[8]

Wilson, hoping for a better result, in 1981 hired a science communication firm to help him design a way to inform oceanographers as to what satellite technology, so new to them, could provide. He felt he had to have science support to persuade NASA’s new leadership to adopt TOPEX. While TOPEX would not be as “big science” as Press advised, it would still be bigger than Seasat. It would cost many hundreds of millions.

Getting a Recommitment

In early 1981, as Wilson and his small band of allies planned the TOPEX project, President Ronald Reagan appointed James Beggs as NASA Administrator. Beggs was a Naval Academy graduate and had served as NASA’s Associate Administrator for Advanced Research and Technology from 1968 to 1969. He then became Undersecretary of Transportation before going to lead General Dynamics, a large aerospace contractor, prior to his return to NASA. Beggs had attitudes favorable to the ocean program not only because of his Naval background but environmental concerns. However, his overwhelming priority was to move the shuttle from development to operations and get from Reagan a decision to build a large space station. Getting a Space Station decision was essential in his view to keep NASA a major agency and enable it to take “the next logical step” in human space exploration.[9]

Wilson had to obtain from Beggs a decision to recommit the agency to an ocean from space program generally, and TOPEX particularly. Wilson now had a new OSSA Associate Administrator, Burton Edelson, who had been Beggs' roommate at Annapolis and was close to Beggs. Edelson wanted to build a big environmental and Earth Science program. An engineer and technical manager, Edelson inherited the ozone program, with its legislative base, and the emerging ocean program, which needed a decision from Beggs to continue.[10]

In February, just prior to Edelson's arrival, NASA's Deputy Administrator, Hans Mark, had met with Richard Goody, a Harvard atmospheric physicist and Michael McElroy, an engineer, now Goddard's Deputy Director. McElroy had been involved in Mars observations and wondered about the possibilities of studying the Earth holistically the way NASA did distant planets. Goody was worried about the "CO2 problem" and climate change. He believed NASA should organize a global observation program to study this issue. Mark encouraged Goody to define an initiative NASA could take to a U.N.-sponsored space conference in Vienna in August (UNISPACE-1982). This he did.[11]

When he came aboard NASA, Edelson embraced Goody's idea of a global climate initiative. Beggs was favorable as long as it did not threaten his Space Station strategy. The initiative evolved into "global change," something broader and more inclusive than climate change. Oceans were part of the potential activity. While Wilson's program was further along in development than the global change possibility, he still needed the NASA Administrator to endorse what he was doing to continue.

On July 13, Beggs invited Wilson to a meeting he was holding with other agency heads with ocean interests. This was an informal gathering Beggs called the "Ocean Principals." He wanted Wilson to brief them (and himself) on what NASA had underway. Wilson gave an explanation of the value of studying the oceans from space. Everyone was enthusiastic about what they heard. Beggs made it clear he was interested, but also explained that NASA was the sole funding agency for space science. Other agencies had ocean concerns. He declared that NASA would be willing to work with sister agencies on an ocean program, but that it would not do so by itself.

Wilson understood he had to find a partner or partners to get Beggs to say "yes." But Beggs obviously was intrigued.[12]

Finding a Partner

Wilson first went to NOAA, which said it was not interested. He then went to the Navy, which explained that it was fully engaged in its own program, a follow-up to the terminated NOSS program (like NOSS it would wind up terminated). Wilson realized that his best hope was across the Atlantic. He knew from his earlier trip that Europeans had interests in the ocean from space.

The issue was how strong their interest was. Would they put money behind their words? Wilson went to space officials in England first and was told they could not contribute except managerially. The Germans said they had other priorities. He now tried the French. He spoke with Lefebvre. The Frenchman told him CNES was not only interested, but planning its own mission to study the ocean from space, one called Poseidon. He explained the French "would welcome the idea of collaborating." Since Lefebvre worked at a field center of CNES, similar to JPL, Wilson had to talk with his CNES Headquarters counterpart in Paris, and he did so. This was Jean-Louis Fellous, Program Manager for Aerospace and Ocean Research.

The Europeans had created an international organization, the European Space Agency (ESA), so they could participate in large space programs requiring joint funding, such as the space shuttle and the biggest rockets. But individual nations had their own space programs for smaller and intermediate-scale activities. The French saw ocean observation as a possible national niche. Wilson returned to Washington, feeling his chances for a Beggs decision were more favorable.[13]

The Vienna Debacle

While Wilson was exploring opportunities for collaborative research, Edelson, advised by Shelby Tilford, a physical chemist, head of NASA's ozone program, was developing the global change initiative. In August 1982, Beggs presented this initiative, called "Global Habitability." As Leshner and Hogan described the reaction of those attending the Vienna conference:

> The response to the Beggs initiative was unambiguously negative for several reasons. Many within the American scientific community felt that they had not been appropriately consulted about the contours of such an undertaking.

> One participant contended that the proposal 'was presented without warning [and as a result] it came across like NASA was trying to take over the world.' Non-American attendees were 'insulted at the implied condescension and worse, members of the international science bureaucracy saw Global Habitability as undermining…existing cooperative programs.' Domestic agencies were not pleased because they thought this program was a NASA power grab intended to diminish their ongoing remote sensing activities.[14]

Back in Washington, Beggs told Edelson, who had been "stunned by this reaction," not to give up but build a constituency before surfacing such a flagship initiative again. The idea of a major program covering all aspects of global change (including climate change) remained on the NASA agenda. It advanced in parallel with the ocean observation program. The difference was that the "Earth Observation System" (EOS), as it would be called in 1994, was conceived as "big science," whereas the ocean program was best characterized as "intermediate science."

Scale mattered to NASA, not just technically, but institutionally and politically—within NASA and externally to NASA. The agency had large field organizations to support and needed big programs to fund them. Big programs also meant large external constituencies. However, the bigger the program, the bigger had to be the coalition to sell it. Edelson told his OSSA organization: "I want you to take a look at what you can do for Earth science with a big platform and polar orbit and what technology would drive the science."

On August 31, in a memo giving guidance for an internal study, Edelson made it clear what he wanted:

> Please aim at a satellite system design for all civilian remote sensing R&D purposes: science—geology, hydrology, ecology, atmospheric chemistry, climatology, and oceanography and potential —agricultural assessment, renewable resource monitoring, ocean monitoring, mapping, meteorology, etc.[15]

To help win Beggs' support, Edelson connected EOS to the Space Station. Beggs wanted the Space Station to have a science relation and Edelson gave that to him. The linkage would help promote the station to the scientific community. It would also lift the Earth Sciences from the shadow of the planetary sciences side of OSSA. It "would give the Earth Science community its first flagship mission."[16]

Edelson thought big. Those Earth scientists at NASA who believed in the Edelson vision were excited by the prospect of linkage with human space flight and the larger budget and prominence they would have.

One who was not enthusiastic was Wilson. What he needed was for Beggs to commit NASA to a joint venture with CNES. His efforts in this respect were hampered by an Edelson reorganization. In April 1983, Edelson appointed Tilford director of a new Earth Science and Division. Tilford had been leading EOS planning along with the ozone depletion program. He and Wilson had been co-equals, organizationally, before the change. Now Tilford was Wilson's boss. Tilford shared Edelson's "comprehensive" vision of EOS, and his new status gave him much influence within NASA. Tilford wanted TOPEX or some instruments thereof to be part of EOS. He also wanted to protect funds for his ozone-monitoring satellite called the Upper Atmosphere Research Satellite (UARS), for which he had already received a decision to build.

An End-Run to a Beggs Decision

How was Wilson to get to Beggs for an "independent" ocean program? A possible opening came in the spring 1983. He learned that a high-level delegation from CNES was coming to visit with Beggs. The delegation included Hubert Curien and Fredrick D'Allest, the President and Director General, respectively, of CNES. Wilson asked Lefebvre if he could help get CNES officials to raise the topic of NASA/CNES collaboration in satellite altimetry in their meeting with Beggs. Lefebvre made the arrangement.

When Beggs saw the possible NASA/CNES collaboration on his agenda, he asked Wilson to attend "and there was no way Tilford could stand in the way." At the meeting, Curien proposed CNES provide a satellite, NASA the launch vehicle, and both countries put sensors on the satellite. Beggs asked Wilson what he thought. Wilson said the French satellite could not meet NASA's requirements for a non-sun-synchronous orbit and therefore was not acceptable to NASA. If CNES would not agree to an alternative, a joint project could not proceed. Well, asked Beggs, was there an alternative NASA could suggest?

Wilson replied that the United States had satellites capable of non-sun-synchronous orbits and the French had a launch vehicle, called Ariane. NASA could provide the satellite and CNES the rocket, a reverse division of labor. D'Allest's eyes almost popped out of his head, Wilson recalled. Wilson did not know it at the time, but the use of Ariane was more than the French could have hoped for as D'Allest was regarded as "the father of Ariane."

Beggs then returned to Curien. What do you think? he asked: Discussion ensued. The French and Beggs agreed to Wilson's proposal. They further said they would meet again in June at the Paris Air Show and decide whether or not to formally commit to a joint program.

After the meeting, further conversations within NASA took place. The big issue was Ariane. Wasn't the United States competing with the French on launch vehicles? Why help a rival? Wilson responded that Ariane was a reality and was not going away. It would make sense, Wilson said, to work with the French, when it was in America's interest to do so.

Beggs thus agreed to the Wilson option. The United States got the satellite and orbit it needed for its priority—the test of an advanced altimeter. The French got credibility for Ariane they craved. Both countries saved hundreds of millions of dollars by collaboration. This marked a precedent: the first time the United States agreed to use a foreign rocket for a major satellite.

In June 1983 at the Paris Air Show, Beggs and Curien made the decision official.[17] There would be further discussions about technical matters, and the White House and Congress would have to acquiesce, but this agreement was the enabling decision for Wilson. NASA would put money into its proposed budget for the mission and CNES would do the same. What Wilson had pushed, NASA as an agency would now promote. The major question was when it would move this initiative for formal presidential and congressional adoption and schedule a launch. NASA had many programs competing for "New Starts" and a constrained overall budget.

Oceanographic Resistance

There were a number of potential new starts coming up for inclusion in NASA's budget proposal in 1984, and some would get priority and others would be held back. One of the factors that weakened ocean advocates' case was the lack of a pressure from the oceanographic community. Wunsch was an active advocate at NASA in the mid-1980s when TOPEX/Poseidon (the name of the merged program) came up for decision. He was deeply involved in WOCE planning, and he wanted the deployment of satellite altimetry to coincide with advent of this international project. WOCE would be funded in part by the National Science Foundation (NSF), and Wunsch was aggressive on that front.

But he had a problem. He "was pointedly asked by the Administrator why there were so few people from the Woods Hole Oceanographic Institute and Scripps Institution of Oceanography involved." Beggs expected much more pressure from the ocean community than he received. After all, planetary scientists assiduously asserted their claims. Wunsch did his best, but the reality was indifference and even resistance on the part of most ocean scientists. Oceanographers were a "macho, individualistic" community in Wilson's view, and they loved their ships.

Wilson had to enlarge his constituency to make TOPEX/Poseidon a priority in the line of potential new starts.[18] He moved to get Scripps aboard by making it a data handling facility. Nierenberg would have been an ally. However, he retired before Wilson could consummate this strategy. Nierenberg's successor, Edward Frieman, wanted the role, but Scripps was connected with the University of California at San Diego, and a faculty committee rejected the assignment. As JPL historian Erik Conway explained it, the academics "had seen a zero-sum game—funds put towards a satellite center would reduce those available to more traditional oceanographic research." That meant JPL stayed fully in charge of ocean satellite data processing, and Wilson lost a potential source of scientific and political support.[19]

Seeking Help

Wilson enlisted the help of D. James Baker, who headed the Joint Oceanographic Institute Inc. (JOI). This was a consortium of ten major institutions in the United States with substantial oceanographic interests. It had been set up to manage an international ocean drilling program. It also was an advocate for oceanographic interests to policymakers in general.

Wilson asked JOI to help in formulating a long-term (beyond TOPEX/Poseidon) program plan and help to get support from ocean scientists and policymakers. JOI formed a satellite planning committee of scientists with expertise in remote sensing and in oceanography. Wilson, Townsend, and Bill Patzert worked closely with them. Patzert was a JPL oceanographer on an extended assignment with Headquarters. He was slated to be Program Scientist for TOPEX/Poseidon along with Townsend as Program Manager. JOI came up with a program plan for four satellite missions, leading with TOPEX/Poseidon. It produced a report, published in 1984, entitled: "Oceanography from Space—A Research Strategy for the Decade 1985-1995."

JOI circulated a summary of the report widely to the oceanography community with the full report as technical back-up distributed more narrowly. Payson Stevens, head of a space-media firm Wilson had previously employed, contributed visual designs that dramatized space-ocean linkages.

In addition to the report, Baker advised Wilson on broader ways to enlist oceanography interests, such as articles about the program in journals ocean scientists read. Baker also worked with Wilson on congressional strategy, providing readable materials as well as briefings to congressional staff. Baker sought out senators with environmental interests, such as John McCain (R., Arizona) and Al Gore (D., Tenn.). Eventually, Congress would have to agree to TOPEX/Poseidon and possibly the program as a whole.[20]

There was no great effort to highlight climate change and sea-level rise in this constituency-building effort. Wilson's team, including Wunsch, did not expect TOPEX/Poseidon to be able to detect that phenomenon of sea-level rise. However, some of the scientists working with Wilson on plans for TOPEX/Poseidon speculated about possibilities, even publishing a scientific article in 1986.[21] A graduate student of one of the scientists, Steve Nerem, decided to continue this avenue via his dissertation and later work. But most researchers associated with the project did not have sea-level rise on their agenda.

Wunsch's Legerdemain

Wunsch continued talking with Beggs and other top NASA officials to try to get a decision. He joined Baker in lobbying George Keyworth, President Reagan's Science Adviser, to get White House backing. In spite of these efforts, progress was slow.[22] TOPEX/Poseidon was passed over in 1984 for the FY 1985 budget and in 1985 for FY 1986. Wunsch wore two hats—a science planner for TOPEX/Poseidon and one for WOCE. He worked to get the two activities to proceed in tandem. His strategy was to tell oceanographic funding agencies (such as NSF) that WOCE had to be within a time frame interval to take advantage of the independently funded satellite mission, while simultaneously telling the space agencies (NASA and CNES) that the satellite had to be flown in a finite time window to take advantage of the independently funded in situ WOCE program. He did not tell NASA that a number of his co-WOCE planners regarded NASA's contribution "as a colossal waste of money." He artfully concealed negative oceanographer attitudes from NASA.[23]

Getting a "New Start"

Whatever Wilson, Baker, and Wunsch did, they could not achieve a "new start" to build TOPEX/Poseidon in 1984 or 1985. Circumstances, however, opened a window for adoption. Thinking in NASA had advanced to a decision to launch a Mission to Planet Earth. That was the title of an editorial by Edelson published in *Science* magazine, in January 1985, announcing that fact. Avoiding the approach of "Global Habitability," he emphasized it was an interdisciplinary, interagency, international initiative.

Since 1983, NASA had been assiduously planning the mission and had established an Earth System Science Committee under an eminent scientist, Frances Bretherton, to help in the planning. After two years, the Bretherton Committee was providing NASA with an intellectual framework that would cut across sciences, agencies, and nations. It called for study of the Earth as a system—"Earth System Science." The editorial by Edelson pointed out that humanity was responsible for some of the negative changes to the planet, including those due to the "extraction of energy from fossil fuels." It now had the technology and the incentive to move forward on this Mission to Planet Earth.

Having included stakeholders in its planning, NASA found the reaction to the editorial as positive. There was a sense that the time was ripe. It was helpful to ocean science advocates, in the sense that 70% of the "Earth System" consisted of the oceans.[24]

An event that helped the ocean advocates' cause came just a few months later. In the May 1985 issue of *Nature* magazine, a British team led by Joseph Farman published a peer-reviewed article about a key element of the Earth system—the ozone layer. Farman had found that the ozone layer blanketing Antarctica had shrunk by 30% in recent years. Ozone depletion had been a problem in the 1970s causing Congress to give NASA the mission to diagnose what was happening. What had been a problem then was considered a crisis now. Ozone protected people from harmful aspects of the sun's rays.

The ozone issue exploded into a huge national and international debate. NASA had missed the ozone "hole" over Antarctica but now took charge of defining the issue from a science point of view. It had the legislative mandate and technology. Satellites showed the scale of the hole—the size of the continental United States. A NASA official under Edelson, Jack Kaye, remarked, "You could watch it grow, evolve like a living organism." The pictures on media were ominous. NASA involved NOAA and the private sector in researching the major cause—chlorofluorocarbon chemicals. NASA scientists also advised policymakers in the United States and abroad, helping them toward the decision to ban CFCs in the Montreal Protocol in 1987.[25] The ozone issue showed the importance of a Mission to Planet Earth. While it aided NASA's ozone program—and UARS—most, it also indirectly helped the ocean advocates with their case.

But the most influential factor triggering a new start launch decision for Wilson was a terrible event—the shuttle Challenger explosion of January 28, 1986. It killed seven astronauts including the first "teacher in space," Christa McAuliffe. The disaster set NASA back in many ways, most significantly its use of the shuttle. It was not until September 29, 1988, that the shuttle returned to flight. Challenger meant that the use of the shuttle would be sharply limited. If there was any other way to get important payloads into space without a shuttle, that was considered a plus in getting a new start launch decision. Obviously, the fact that TOPEX/Poseidon would be launched by Ariane was a positive argument.[26]

These three factors converged in 1986: Edelson's desire to go full speed with a Mission to Planet Earth, the ozone hole, and the shuttle disaster. They expedited NASA's decision to launch TOPEX/Poseidon. UARS would go up on a shuttle in 1991; TOPEX/Poseidon would follow in 1992 on Ariane. NASA's MTPE would begin with these two precursory missions.

NASA put TOPEX/Poseidon into its FY 1987 budget. The White House and Congress agreed. The French followed suit. Indicative of its priority for France, the French decision was announced triumphantly by Jacques Chiarac, the French president. He was accompanied in the announcement by Alain Madelin, Minister of Industry and Jacques Valado, Minister of Research. The French were very much a part of Wilson's advocacy coalition.[27]

Notes

1. Dixon M. Butler, interview by Rebecca Wright, Oral History, EOS Collection, NASA, June 25, 2009. Retrieved from shorturl.at/tBCHJ.

2. W. Stanley Wilson, "Why Would NASA Collaborate with CNES in Altimetry?," *History of Meteorology, Atmosphere, and Ocean Science from Space in France and Europe by its Actors* Ed. by Fellous, Jean-Lewis., (Paris, France: Institut Francais d'Histoire de l'Espace, forthcoming).
3. Bob Winokur, interview by author, Sept. 21, 2021; W. Stanley Wilson et al. "Satellite Oceanography—History and Introductory Concepts," in *Encyclopedia of Ocean Sciences*, 3rd Edition, vol. 5, (2019), 347–361.
4. Carl Wunsch, interview by author, Sept. 13, 2020.
5. B.J. Thompson, et al. "The Origins, Development and Conduct of WOCE," *International Geophysics* vol. 77, 2001, 31–43, 1-viii.
6. W. Stanley Wilson, "Why Would NASA Collaborate with CNES in Altimetry?" *History of Meteorology, Atmosphere, and Ocean Science from Space in France and Europe by its Actors* Ed. by Fellous, Jean-Lewis., (Paris, France: Institut Francais d'Histoire de l'Espace, forthcoming).
7. Robert Stewart, "Early Days of TOPEX at the Jet Propulsion Laboratory," in Fellous.
8. Carl Wunsch, "TOPEX/Poseidon as a Science Mission" in Fellous; See also Carl Wunsch, "Towards the World Ocean Circulation Experiment and a Bit of Aftermath," in M. Jochum and R. Murtugudde (Eds). *Physical Oceanography*, (NY: Springer, 2008), 181–201. Retrieved from http://link.springer.com/10.1007/0387-33152-2
9. Howard McCurdy, *The Space Station Decision: Incremental Politics and Technological Choice* (Baltimore, MD: Johns Hopkins, 1990).
10. Richard Leshner and Thor Hogan, *The View from Space* (Lawrence, Kansas: University of Kansas Press, 2019).
11. Ibid., 38.
12. W. Stanley Wilson, interview by author, Oct. 27, 2020.
13. Ibid.; See also W. Stanley Wilson, "Why Would NASA Collaborate with CNES in Altimetry?," *History of Meteorology, Atmosphere, and Ocean Science from Space in France and Europe by its Actors* Ed. by Fellous, Jean-Lewis., (Paris, France: Institut Francais d'Histoire de l'Espace, forthcoming).
14. Richard Leshner and Thor Hogan, *The View from Space* (Lawrence, Kansas: University of Kansas Press, 2019), 61–62.
15. Ibid., 63.
16. Ibid., 198.
17. W. Stanley Wilson, "Why Would NASA Collaborate with CNES in Altimetry?," *History of Meteorology, Atmosphere, and Ocean Science from Space in France and Europe by its Actors* Ed. by Fellous, Jean-Lewis., (Paris, France: Institut Francais d'Histoire de l'Espace, forthcoming).
18. Carl Wunsch, "TOPEX/Poseidon as a Science Mission," in *History of Meteorology, Atmosphere, and Ocean Sciences from Space in France and*

Europe by its Actors, edited by Fellous, Jean-Lewis., (Paris, France: Institut Francais d'Histoire de l'Espace, forthcoming).

19. Erik Conway, "Drowning in data: Satellite Oceanography and Information Overload in the Earth Sciences," in *Historical Studies in the Physical and Biological Sciences* vol. 37, No. 1 (Sept, 2006), 147.
20. D. James Baker, "Oceanography from Space—Building a Community Consensus," in Fellous; See also James Baker, interview by author, Sept. 9, 2021.
21. GH Born et al., "Accurate Measurement of Mean Sea Level Changes by Altimetric Satellites," *Journal of Geophysical Research,* (1986), 91:11,775–11,782.
22. Lee-Lueng Fu, Interview by author, Sept. 24, 2020.
23. Carl Wunsch, "TOPEX/Poseidon as Science Mission," in *History of Meteorology, Atmosphere, and Ocean Sciences from Space in France and Europe by its Actors,* edited by Fellous, Jean-Lewis., (Paris, France: Institut Francais d'Histoire de l'Espace, forthcoming).
24. Richard Lesner and Thor Hogan, *The View from Space* (Lawrence, Kansas: University of Kansas Press, 2019).
25. W. Henry Lambright, *NASA and the Environment: The Case of Ozone Depletion* (Washington, DC: NASA, 2005).
26. Charles Yamarone, interview by author, Nov. 20, 2019.
27. Jean-Louis Fellous, "The Difficult Beginning of a Perfect Agreement," manuscript sent to the author, also in Fellous, ed.

CHAPTER 4

Breakthrough for Sea-Level Rise

Abstract Chapter 4 covers the 1992–1994 period. TOPEX/Poseidon was not expected to be able to measure sea-level rise. A prime goal was to study large currents and how they might impact climate. But the satellite resolution was so precise, many scientists realized that they could detect with clarity climate change's impact on the oceans. In 1994, a NASA-based scientist found a way to translate complex data into sea-level measurements. Doing so was a breakthrough and it attracted not only scientific but media attention. Meanwhile, NASA Administrator Dan Goldin forced out Tilford and proclaimed EOS the kind of "big science" that conflicted with his faster, better, cheaper (FBC) mantra. TOPEX/Poseidon was a success from the Goldin FBC perspective.

Keywords Daniel (Dan) Goldin • Shelby Tilford • Big science • Faster better cheaper (FBC)

On March 23, 1987, NASA and CNES signed a Memorandum of Understanding to develop TOPEX/POSEIDON. The MOU stated its primary mission was to measure surface topography globally for at least three years in a way to permit the determination of global circulation and its role in climate. The MOU made it clear that NASA and CNES wanted the mission to lay the foundation for a continuing program of long-term

W. H. Lambright, *NASA and the Politics of Climate Research*, Palgrave Studies in the History of Science and Technology, https://doi.org/10.1007/978-3-031-40363-7_4

observation of ocean circulation and its variability. The agreement mentioned the connection of the NASA–CNES venture with other international programs, including WOCE.

While the agreement was significant, what really mattered to NASA leaders was the ocean program's precursory connection to EOS. EOS was where NASA's Earth Science priority lay. What happened to TOPEX/Poseidon affected what happened to EOS. Their futures were intertwined. Over the next five years, 1987–1992, NASA and CNES worked out cultural and other differences and developed TOPEX/Poseidon hardware. NASA also forged ahead promoting EOS as a top agency priority. EOS initially was an asset to TOPEX/Poseidon, later it became a threat. At the same time, climate change grew as an elephantine issue in the background influencing both endeavors. While those who worried about climate change raised the danger of sea-level rise, very few believed TOPEX/Poseidon or EOS would clearly detect it.

Developing Technology and Partnership

Once policymakers decided to authorize and fund TOPEX/Poseidon, technical teams began working out details of the project. They also evolved a partnership that expedited the development. They had to compromise on many small and large differences, including when to start meetings, where to meet, project name, language, expenses, and critical technical matters such as orbits and data access.

One of the most difficult questions was technology transfer. Townsend, a chief negotiator, took a hard line to protect U.S. interests. Wilson, knowing that if the two sides could not agree, the project would die, thought his associate was too inflexible. The two men took long walks in Paris, where a number of meetings were held. They talked extensively, sometimes over drinks. In the end, NASA and CNES evolved a "black box" strategy. They would discuss particular components of TOPEX/Poseidon, but not what was inside those components.[1]

Both teams of negotiators wanted to keep what they regarded as "technical" decisions at their level, fearing escalation up to higher echelons that might make their choices about development impossible. They largely succeeded and found they could work as partners, avoiding questions of who was in charge of what as much as possible. What mattered was to get TOPEX/Poseidon built!

Post-Challenger: EOS Rises

As technical development commenced, decisions evolved simultaneously at policy levels. NASA leadership had changed by 1987. Beggs had left NASA in December 1985, before the Challenger accident, to fight what were proved to be false allegations of criminal conduct before joining NASA when he was at General Dynamics. President Reagan prevailed on James Fletcher to return to NASA in May 1986 to direct the agency's recovery after the disaster and damning investigation. In July 1987 Edelson retired, to be replaced by Lennard Fisk, a physicist and former professor at the University of New Hampshire. Fisk had been involved with the Bretherton study and firmly believed that NASA's MTPE and Earth Observation System were the right priorities. Tilford remained in his pivotal position leading MTPE and EOS. EOS was conceived as consisting of two huge satellites, providing *comprehensive* views and *simultaneous* measurements of all aspects of an Earth System Science, including the oceans. He viewed the concept of comprehensiveness and simultaneity critical to NASA's plans and rhetorical advocacy for EOS.

While Fletcher gave most of his attention to human space flight—shuttle return and Space Station design—he backed MTPE and its EOS ambitions. He wanted NASA—and the nation—to look forward, not behind at Challenger. Toward that end, he asked Sally Ride, the first American woman in space, to lead a study of what should be the agency's top priorities for the future. Her report, published in August 1987, listed four. The first on her list was "Mission to Planet Earth."[2]

A Sense of Momentum

Never before had the Earth sciences felt such a wind of support. After the ozone hole, there was a new sense of environmental momentum. It was not just domestic, but international, and climate change was coming to the fore.

NASA's most outspoken scientist on climate change was under Fisk and his directorate, the Office of Space Science and Applications. He was James Hansen, a senior climate scientist and director of the Goddard Institute for Space Studies in New York City.

In June 1988, Hansen appeared before the Senate Committee on the Environment and Natural Resources and declared: "The Greenhouse Effect has been detected and it is changing our climate now…we already reached the point where the Greenhouse Effect is important." His words came at a point when the American Southwest was experiencing an extremely hot summer and the American Midwest a long drought. Media coverage linked the weather effects with climate change and Hansen's testimony and subsequent comments to the media seemed to reinforce and lend legitimacy to that view.

"It's time to stop waffling," Hansen declared, "and to say that the Greenhouse Effect is here and it is affecting our climate now."[3] Several legislators and environmentalists wanted to see policy action, but many scientists were skeptical. There was a groundswell in many quarters to find out what was happening. The United Nations Environmental Program (UNEP) and World Meteorological Organization (WMO) established an Intergovernmental Panel on Climate Change (IPCC), an informal assembly of climate scientists from many nations to assess what was known and report to policymakers.

The overriding sense of the times was a need for more research to clarify the facts. That sense worked in favor of what NASA was proposing with EOS. NASA, NOAA, NSF, and USGS began to coordinate their research programs so as to get answers. Vice President George H.W. Bush, campaigning for president in 1988, joined the chorus and said he would lead the drive on climate change.

As the prospects for NASA's MTPE grew, so did those for EOS adoption. When Bush was elected president in November 1988, there was an expectation that what was a NASA priority was going to become a national priority. Edelson was gone, but his legacy, EOS, was coming into political favor. He had chosen the name EOS in 1994 because it was the name of a Greek Goddess of the Dawn. He had said it would mark the dawn of a new era in Earth Science and remote sensing. That seemed to be happening.[4]

Bush's Priorities

In May 1989 President Bush used the Apollo landing anniversary to make a major space policy address. He wanted to use space to galvanize American science and technology in place of the Cold War, which was ending. He proclaimed that NASA would return to the Moon and go on to Mars. At the same time, he told NASA to proceed with its MTPE. For NASA's Earth scientists, this was a signal of unprecedented White House support.[5]

One of NASA's first actions was to separate EOS from the Space Station. Technically, the union never made much sense. Politically, Bush's speech and subsequent decision to make global change an official presidential priority indicated the programs no longer needed one another.

Tilford, backed by Fisk and Bush's NASA Administrator, Richard Truly, pushed for EOS adoption. EOS, after years of planning, was proposed as three sets of satellites. Each set of two would go up for a minimum of five years. One set would be followed by another, and then another, for a 15-year period overall. The satellites would be huge—each the size of the Hubble Telescope—and carry as many sensors as possible. NASA's drumbeat was "comprehensiveness and simultaneity." EOS would be the technological embodiment of the Earth System Science concept of NASA and the Bretherton committee. The cost would be $17 billion for the first decade, and many billions more beyond. Some projections went as high as $50 billion over an extended lifetime, a prospect making EOS potentially the government's premier big science program.[6]

The president proclaimed EOS as the space element of an interagency U.S. Global Change Research Program (USGCRP). Global change was about more than climate change, but climate change was the core. EOS would include ocean observations essential to an integrated Earth system approach. In the short run, that grand vision made TOPEX/Poseidon success as a precursor for EOS a priority. In the long run, "comprehensiveness" was a possible threat to the autonomy and separate identity of the ocean program.

While the White House was in the lead in 1989, Congress was amenable, even enthusiastic. It was a heady time for many connected with EOS. Dixon Butler, one of Tilford's chief EOS planners, recalled his feeling at this time:

> We had a sense of imperative because we could see that environmental change was happening, and we knew we needed to go understand it better, and we knew we needed to get our hands around it, and we knew that eventually people were going to need to make informed policy decisions, based on what we were doing. People in the Earth Observation System, on all the working groups, and all the people at NASA—we woke up and knew we were engaged in trying to save the Earth. It's a little bit of hubris, but it was very motivated.[7]

EOS vs. "The Ocean from Space"

In 1990 the president and Congress officially adopted EOS. Funds were put into NASA's FY 1991 budget to get started. The funds were modest, but expectations were that they would ramp up as hardware development commenced. Also in 1990, Congress enacted legislation to start the U.S. Global Change Research Program, the interagency effort of which NASA's EOS was by far the largest part. Congress authorized agencies to set up major research programs, but also called for "usable science." It wanted science to aid in policy.

While Wilson cooperated on EOS planning, he was also critical of its design. What was good for EOS technically was sub-optimal for ocean altimetry in his view. Moreover, he and Tilford did not get along. Butler complained that the ocean unit wished to keep to itself.[8] Wunsch was even more outspoken as a critic of EOS than Wilson and paid a price. He found himself "persona non grata" with Tilford, whom he regarded as more a politician than scientist.[9] Tilford pushed hard for his "big science" approach and needed a united scientific front to get the massive resources required for EOS. Wunsch refused to toe the line. He recalled: "My relationship with NASA went downhill when Tilford was pushing EOS. I opposed his big platform idea. It was bad for science, bad for altimetry. I was vocal.... People [at NASA] wouldn't talk to me."[10]

As EOS's star rose, integration became the guiding organizational philosophy for the program. Wilson's ocean program was abolished as an independent branch of OSSA. Wilson lost his considerable influence and became a program scientist for EOS. Physical oceanography still existed but did not have the autonomy and status, and Wilson the power, that they had previously enjoyed. EOS was "the only game in town," commented Jay Zwally, a NASA glaciologist pushing for inclusion in EOS of ice observation.[11]

Tilford and Wilson both had technocratic visions for NASA's Earth-oriented mission, but their visions clashed. One of Wilson's associates remarked that as TOPEX/Poseidon moved closer to culmination, it was "poignant" to see Wilson pushed aside.[12]

EOS Under Fire

Yet EOS's fate quickly changed. As growing national interest in climate change had expedited EOS adoption in 1990, so larger politics—White House-Congressional decision making—worked the opposite way in 1991. Jack Fellows, who worked in the White House Office of Management and Budget (OMB) on EOS's budget, remarked: "Between FY 1991 and FY 1992 the world changed."[13] What he meant was that the economy began to sour and the federal deficit escalated. Bush had to renege on his campaign pledge not to raise taxes, and he and Congress agreed on budget caps. These caps meant that NASA could not afford to ramp up EOS *and* the Space Station *and* Bush's Moon/Mars program all at the same time. Something had to give.

Congress threatened to kill the Space Station, forcing Bush to choose. He made the Space Station his number one priority. The Moon/Mars program essentially went away. And both president and Congress decided that EOS needed to continue but be considerably downsized. In 1991, Congress told NASA that instead of $17 billion in its first 10 years, EOS would have to make do with $11 billion. The White House appointed an independent engineering review committee to help determine how that would be done.

The answer came back that EOS could not be as comprehensive and should focus more on climate change. Also, it could accomplish what it wanted to do with "simultaneity" through a cluster of smaller satellites flying in formation. Hansen broke ranks with Tilford and told the panel that climate change could not wait for EOS. He proposed a smaller satellite called "CLIMSAT" that should be built much sooner.[14] Tilford went to the White House to plead his case, but a presidential aide told him EOS would take too long to build to do the Administration much good.[15]

Wilson Departs

Meanwhile, in early 1992, Wilson departed NASA. "I was fed up," he recalled.[16] An opportunity for a higher-level position in NOAA opened and he took it. The position was operational, not one of research and development. He no doubt had hoped to stay through the launch of TOPEX/Poseidon, scheduled later that year. But he decided that was not possible. He moved on. Meanwhile, TOPEX/Poseidon was now in its final stage of development and testing.

Close Calls

Six months before the scheduled August launch, the ocean program received a serious scare. The satellite was being tested at the Goddard Space Flight Center in Maryland. It was supported by a crane held high near the ceiling. Suddenly, the crane snapped. The $700 million satellite dangled precariously, fastened only by a wire cable. Fortunately, there were technicians present. They moved rapidly and secured the precious machine at the last minute.[17]

Then, just two months before launch, NASA suffered another potential show-stopper. Some of the batteries essential to powering the spacecraft showed evidence of defects. Would they last through the project's three- to five-year mission? Fisk called an emergency meeting of TOPEX/Poseidon officials to decide whether to delay the launch and develop a substitute power source. The OSSA Associate Administrator knew such a decision had significant cost implications. It would be risky to proceed, but how risky? Engineers believed the batteries could survive at least the three-year design life of the satellite, if not the desired five-year extended life. Fisk decided to take the risk and maintain the schedule, telling the team to treat the batteries with "tender loving care."[18]

The Launch

As the actual day of the launch approached, Dan Goldin, who had replaced Bush's appointee, Richard Truly, as NASA Administrator in April, met with NASA officials most closely connected with the project. With Wilson gone, Patzert and Townsend were providing leadership under Fisk and Tilford. Fisk told Patzert to stay behind in Washington and go with Goldin to the French embassy for a celebration, assuming a successful launch. Goldin, however, instructed Patzert: "Go to the launch and have some fun." Patzert tried to explain that he had to stay behind, but Goldin said again—in front of Fisk—"go to the launch and have some fun."

As it turned out, most NASA officials who went to the Kourou, French Guiana launch site went by commercial aircraft. CNES leaders used a French plane they had chartered for the flight and invited Patzert to come along in their plane. Patzert found a way to get Wilson aboard. CNES invited him, and he accepted.[19]

On August 10, the day of the launch arrived. Patzert later recalled his feelings at the time of launch. "NASA and CNES had labored for a decade to design, fund, and build this beautiful bird." It was now ready, and "when the sun was setting, the rockets fired, we were all on pins and needles. Wow, there was a lot of tension and excitement that evening!" A few more anxious moments passed and the spacecraft thundered into the atmosphere. TOPEX/Poseidon soared, on its way to an orbit 830 miles above Earth.

Patzert remembered the mood of those watching immediately changed from anxiety to elation. Then, the celebration began and lasted all evening. The night skies changed to dawn. For Patzert, viewing the launch and being part of TOPEX/Poseidon was a "privilege."[20]

A Triumph and Surprise

TOPEX/Poseidon was everything for which its advocates had hoped—and much more. As data were analyzed, it was clear that TOPEX/Poseidon was able to view 90% of the world's Ice-free ocean. It soon relayed information making it possible to create a global map of the seas every 10 days. Never before had such large-scale, global observation been possible. Revealed were ocean circulation and its many currents. It was capable of identifying and tracking massive climate-forcing phenomena such as a Pacific Ocean warming El Nino as it unfolded.

The satellite's altimeter was incredibly accurate, far more so than had been expected. It could take measurements down to 3 centimeters (1.18 inches). As NASA and scientists connected with the project saw what the technology could do, many were surprised, even amazed. Wunsch had not expected a capability so remarkable as to measure sea-level rise.[21] There it was. It would take months, maybe years, of analyzing data and developing credible techniques to measure sea-level rise. But NASA, CNES, and their academic science partners knew it was now possible. The key was the extreme altimeter precision that TOPEX/Poseidon provided. Ironically, the impact of the ocean on climate had been a prime driver of TOPEX/Poseidon, but the impact of climate change on the ocean was now an unexpected and vitally important outcome. It was a breakthrough!

That breakthrough was due in part to a decision Wilson had made to bring geodesy experts into his science and technology team. That choice had been disputed by others, particularly ocean scientists. But it had helped make a difference in the orbit flown that was critical to precision.[22]

Soon after TOPEX/Poseidon was pronounced a success, Fisk told the media that "TOPEX/Poseidon will tell us more about the oceans than all the ships in history." It was "an absolutely essential mission" in the global study of the climate of the Earth, and more importantly, he declared, in learning what humans are doing to that climate. He predicted the new capability would help policy makers "protect the future of the planet."[23] Walter Monk, a leading oceanographer, later called TOPEX/Poseidon "the most successful ocean experiment of all time."[24] Wunsch, who had bet his reputation on the potential of ocean satellites, wrote that TOPEX/Poseidon was an "engineering, scientific, and political miracle."[25] The French agreed. "Several miracles," the CNES leader of the project, Jean-Louis Fellous, declared.[26]

Continuity a Must

TOPEX/Poseidon persuaded oceanographic skeptics that space observation was indeed a revolutionary technology. WOCE was a user and could quickly put the new system to a test. The traditional oceanographic tool, the ship, was still important for intended studies. Satellites had limits in getting at heat below the ocean surface. But satellites provided the large-scale perspective ships and other devices could not. Advocates believed that once sea-level measurements could be perfected, and deployed over several years, there would be proof that climate change was truly underway. Scientists and environmentalists had postulated sea-level rise. TOPEX/Poseidon could validate it.[27]

NASA and CNES knew the ocean satellite program had to continue beyond TOPEX/Poseidon. Advocates held that it was too important to be a handmaiden of EOS. It was essential in its own right. Also, TOPEX/Poseidon was not just a technical success. It was an organizational and international success. The NASA–CNES partnership worked. NASA had paid two-thirds of the costs, and CNES one-third. It would not have happened without partnership.

CNES Director General Jean-David Levy announced in the wake of TOPEX/Poseidon that France and NASA were already discussing collaboration on a successor. Fisk could not commit to a follow-up project, but stated it was "absolutely essential that we continue cooperation of this type."[28] "You've had your first success as NASA Administrator," one of the TOPEX/Poseidon researchers, Charles Yamarone, told Goldin.[29]

Notes

1. W. Stanley Wilson, interview by author, Oct. 19, 2020.
2. Richard Leshner and Thor Hogan, *The View from Space* (Lawrence, Kansas: University of Kansas Press, 2019).
3. Mark Bowen, *Censoring Science: Inside the Political Attack on Dr. James Hansen and the Truth of Global Warming* (NY: Dutton, 2008), 128.
4. Richard Leshner and Thor Hogan, *The View from Space* (Lawrence, Kansas: University of Kansas Press, 2019).
5. Thor Hogan, *Mars Wars* (Washington DC: NASA, 2007).
6. Richard Leshner and Thor Hogan, *The View from Space: NASA's Evolving Struggle to Understand the Home Planet* (Lawrence, Kansas: University of Kansas Press, 2019).
7. Dixon Butler, interview by Jennifer Ross-Nazzal, Oral History, EOS Collection, NASA, March 26, 2010.
8. Dixon Butler, interview by Rebecca Wright, Oral History, EOS Collection, NASA, June 25, 2009.
9. Carl Wunsch, interview by author, Sept. 13, 2020.
10. Ibid.
11. Jay Zwally, interview by author, Sept. 28, 2021.
12. Lee-Lueng Fu, interview by author, Feb. 18, 2020.
13. Richard Leshner and Thor Hogan, *The View from Space* (Lawrence, Kansas: University of Kansas Press, 2019), 131.
14. Ibid., 141.
15. Michael Freilich, interview by author, Feb. 18, 2020.
16. W. Stanley Wilson, interview by author, Oct. 27, 2020.
17. Lee-Lueng Fu, interview by author, Aug. 24, 2020.
18. Jean-Louis Fellous, "The Difficult Beginning of a Perfect Agreement," in *History of Meteorology, Atmosphere, and Ocean Science from Space in France and Europe by its Actors* Ed. by Fellous, Jean-Lewis., (Paris, France: Institut Francais d'Histoire de l'Espace, forthcoming).
19. William Patzert, interview by author, Nov. 12, 2020.
20. "William Patzert," *Solar System Exploration*, (NASA, Jan. 24, 2019). Retrieved from https://solarsystem.nasa.gov/people/1220/William-patzert
21. Carl Wunsch, interview by author, Sept. 13, 2020.
22. Robert Stewart, letter to author, Nov. 27, 2019.
23. Noel McCormack, "Lessons from Space for the Ocean Planet," *Space Times* (Sept.–Oct., 1992).
24. Walter Munk, "The Evolution of Physical Oceanography in the Last Hundred Years," *Oceanography*, vol. 15, (2002), 135–141.

25. Carl Wunsch, "Towards the World Ocean Circulation Experiment and a Bit of Aftermath," in *Physical Oceanography: Development Since 1950*, Edited by Markus Jochum and Raghu Martugudde, (New York, NY: Springer, 2006), 181–201.
26. Jean-Louis Fellous, "The Difficult Beginning of a Perfect Agreement," in Fellous, ed.
27. John R. Williams, "TOPEX/Poseidon: Mission to Planet Earth Turns its Gaze Upon the Oceans," *Final Frontier*, (Oct. 1992).
28. Peter B. de Selding "Poseidon Altimeter Spawns Plan for Ocean-Sensor Fleet," *SpaceNews* (Feb. 8–14, 1993), 8.
29. Charles Yamarone, interview by author, Nov. 20, 2019.

CHAPTER 5

Remaking a Mission

Abstract In the 1994–1998 years EOS was downsized greatly, and the sea-level rise mission gained independence. Goldin brought Charles Kennel aboard to remake the Earth Sciences mission (then called Mission to Planet Earth). Kennel served from 1994 to 1996. His deputy, Bill Townsend, had been Wilson's deputy and was an advocate for a follow-on to TOPEX/Poseidon, Jason. As Jason, known as Jason-1, was carved from a deconstructed EOS so also was another satellite, ICESAT (Ice, Cloud, and land Elevation Satellite), freed to help get at the "whys" of sea-level rise. Kennel responded to increased climate change politics and visibility by a communication strategy that stressed the "facts" but refrained from alarms and advocacy of policy action. When Kennel left, Townsend took over leadership of Mission to Planet Earth from 1996 to 1998 on an acting basis. He added a pair of sea-level-related satellites called GRACE (Gravity Recovery and Climate Experiment) that complemented ICESAT by showing how polar melting transferred to ocean mass.

Keywords Charles Kennel • William (Bill) Townsend • Jason-1 • Ice, Cloud, and land Elevation Satellite (ICESAT) • Mission to Planet Earth • Gravity Recovery and Climate Experiment (GRACE)

W. H. Lambright, *NASA and the Politics of Climate Research*, Palgrave Studies in the History of Science and Technology, https://doi.org/10.1007/978-3-031-40363-7_5

TOPEX/Poseidon was a great success, and advocates on both sides of the Atlantic pressed immediately for a follow-on. The problem was that Wilson was gone and an upheaval at OSSA underway. NASA Administrator Goldin, an engineer and executive from industry, had a clear and dogmatic philosophy when it came to robotic spacecraft whether for distant planets or Earth. It was "faster, better, cheaper." TOPEX/Poseidon was a model for what he wanted—not EOS. At the same time, Goldin did not like "follow-ons." NASA, he insisted, was about new technology, "firsts." Goldin's views mattered, especially since he remained NASA Administrator under Bush's successor, Bill Clinton.

In the remainder of 1992 and the Bush Administration, Goldin reorganized OSSA and removed both Fisk and Tilford. Retained by President Bill Clinton in January 1993, Goldin strengthened his role and rule at NASA. It was not until 1994 that clear leadership returned to MTPE. TOPEX/Poseidon, meanwhile, provided torrents of data, and as a result, sea-level rise findings made this issue come more into scientific focus and media prominence. From 1994 to 1996, Goldin's appointee as MTPE Director, Charles Kennel, restructured MTPE/EOS in line with Goldin's philosophy.

He also formulated a political strategy for NASA's Earth mission to cope with the growing partisan divide in Congress on climate change. He sought a clear boundary between science and politics. Meanwhile, Kennel's deputy, Townsend, provided continuity from the Wilson era and protected sea-level rise as a mission, helping it regain its independent, stand-alone status. When Kennel left, Townsend served as Acting Associate Administrator from 1996 to 1998. He consolidated the changes Kennel and he had made and further bolstered the sea-level mission.

Fisk and Tilford Go

In September 1992, Goldin announced a reorganization of the Office of Space Science and Applications, which sent a shock wave throughout NASA and the scientific community. The change was driven by Goldin's technical philosophy and desire to replace both Fisk and Tilford.

Goldin said NASA, in general, and OSSA, in particular, were afflicted by a "vicious cycle." His view was that:

> NASA loaded a large number of experiments onto a few big, expensive machines that were launched into space. The scale of the enterprise meant

> that it took a long time to get those spacecraft developed and operating. Because it took so long to get the spacecraft built, they incorporated obsolete technology by the time they reached orbit. With so much incorporated into these expensive machines, NASA could not afford to lose any of them. The agency had become risk averse…and emphasized extra reliable (i.e., less innovative) technology. If anything ever did go wrong, NASA took a huge political hit because so much money and time appeared to have been wasted.[1]

Goldin cited EOS as a negative example, calling it a "Battlestar Galactica," a reference to a gigantic spaceship featured on a contemporary television program. Goldin's solution was to take advantage of many microelectronic technical advances to produce smaller spacecraft that cost less and could launch more frequently. He had implemented such a philosophy as a Vice President for Space while at TRW, a large aerospace firm.

In September, he abruptly removed Fisk as OSSA's leader, reassigning him to a new position of Chief Scientist, with no program and budget. With Fisk gone, Goldin split OSSA into various parts, the most significant being an Office of Space Science and a separate MTPE. He made Tilford "Acting" Associate Administrator for MTPE. The emphasis was "Acting." Tilford knew he was on the way out. He and Goldin had crossed swords when Goldin was in industry. As Goldin saw it, Tilford had threatened TRW, his company, with losing contracts in order to silence Goldin's advocacy of a faster, better, cheaper way to build EOS. That was not how Tilford saw it, but now Goldin was his boss. Tilford commented that the first time he walked into Goldin's Administrator office, Goldin "accused me of personally undermining TRW in a selection process." Like Fisk, Tilford left NASA by the end of 1992.[2]

Kennel and Townsend

Al Gore, Vice President as of January 20, 1993, liked Goldin, and hence he stayed as NASA Administrator. He pressed ahead with his reforms for the unmanned programs, giving most of his time—like all NASA administrators—to human-space flight. He searched for a replacement for Tilford in 1993, while Patzert and Townsend held the program together. At the end of 1993, he found the person he wanted.

He was Charles Kennel, a UCLA astrophysicist, not an Earth scientist. Why me? Kennel asked. Goldin responded: "You are going to have to make some very tough decisions about downsizing the Earth Observing

System. It will be easier for you if you don't have to do it for colleagues that you've known for thirty years." What Goldin intended to do was rachet down EOS from big science to an intermediate-scale model, more in line with TOPEX/Poseidon.

There was another reason he chose Kennel. Politically savvy, Goldin saw that climate change was becoming more conflicted, especially with Gore as Vice President and republicans in Congress making it a partisan issue. "This office," he told Kennel, "deals with some of the most controversial issues NASA has...and there is a whole group of people in the Congress who believe that the scientists have cooked up those problems and are crying wolf and they're doing it to feather their own research nests."[3]

He declared that Kennel was "a first-class scientist with an impeccable pedigree...I want the world to know that science is in charge of Mission to Planet Earth." With Gore's personal blessing, Kennel began as Associate Administrator January 6, 1994. To help him with Earth science, he had Robert Harris from the field, as Research Director. His deputy was Townsend, the same individual who had been Wilson's closest associate and was a long-standing ocean altimetry advocate.[4] With Kennel concentrating on keeping as much science as possible in EOS as it went through rescoping and downsizing, Townsend worked to secure the ocean program as part of the new model.

Sea-Level Findings

As Kennel and his team remade EOS, researchers deep in the organization sought to define the sea-level rise signal from the mass of satellite data. To get accurate measurements, TOPEX/Poseidon had to orbit the Earth continually and bounce radar signals off the sea's surface, thereby measuring the "distance between the satellite and the water." It did that. It also had to measure height of the satellite relative to the center of the Earth. The aim was to free the measurements from "distortion by rising or sinking land." There were other actions that had to be taken to remove "the influence of the tides and the waves."[5] With massive data and powerful computers, results came in that scientists could extrapolate. The only question was who would succeed first.

Steven Nerem had started studying sea-level rise possibilities as a doctoral student at the University of Texas at Austin in the late 1980s.He was then part of a Wilson-supported research team. He was now based at

NASA's Goddard Space Flight Center. TOPEX/Poseidon had shown it was possible to detect a sea-level rise signal. The problem was to do it in a scientifically credible way. As Nerem remembered: "I was an aeronautical engineer and interested in probing the limits of satellite accuracy. What could you really see? There was a lot of noise in the data." Then, one day, his hunting paid off. He saw "a clear signal of a rise." Telling a colleague what he had found, the colleague told him: "Steve, publish this," and you will be "Mr. Sea-Level Rise."[6] Nerem posted the findings in an internal Goddard bulletin and in December, 1994, made his findings broadly public through a paper delivered at the American Geophysical Union annual meeting. He said that global sea-level had gone up 3 millimeters, or about one-tenth of an inch a year since TOPEX/Poseidon launched.

This amount did not seem very large, but it was regarded by many as quite significant. Most importantly, it attracted the attention of the media. *The New York Times* relayed the findings, noting that if the trend continued, that would be dire for low-lying coastal cities and countries.[7] There had been lots of talk about sea-level rise, but these measurements were based on hard science. They were significant evidence of climate change.

Nerem emphasized that two years of data were not *conclusive* evidence. But they were in line with what models of climate change predicted. The data entailed 500,000 sea-level measurements over the globe over two years. The measurements were remarkably clear. With all the other data that went into calculations, it was possible to isolate mean sea-level change.

"The important thing" about these findings, Nerem pointed out, was that for the first time, climate modelers were getting data that were precise enough to reveal the subtle signatures of global change. He added: "In another few years, we'll have a pretty good idea of the true rate of sea-level rise, something that interests a lot of people besides scientists."[8]

Other scientists said it was premature to validate sea-level rise. But the data were publicly available for checking. Nerem publications followed, and gradually there was scientific acceptance. Meanwhile, at NASA Headquarters, top officials were reshaping EOS to assure a continuing record.

Reshaping Earth Science

The Earth science community was not initially happy with an astrophysicist directing MTPE. However, Kennel involved many academic Earth scientists in the EOS restructuring. His fundamental decision was that what mattered for science were the sensors on the satellites not the size of the satellites.[9] The Edelson/Tilford approach was to build satellites big enough to incorporate a comprehensive suite of sensors giving simultaneous views.

Kennel regarded twenty-four of these sensors as most critical, and they could be divided and placed on smaller satellites that cost less and could be launched over time. EOS thus came down in size and price. It was redefined into three intermediate-scale satellites costing around $1 billion each that could launch as money was available, not all at once. These would carry a cluster of sensors in a particular area of science, respectively emphasizing land, water, and atmosphere. This first set would be followed eventually by second and third sets of similar scale to provide a long-term record.

By substantially scaling back the revised EOS, there would be money freed for smaller, specialized satellites in particular fields of great importance. Decisions were made to have altimetry not on EOS, but on one of these stand-alone satellites. One was called Jason, a follow-on to TOPEX/Poseidon. Another stand-alone satellite was named ICESAT and would study glacial and ice-sheet melting. Both of these pertained to sea-level rise. Jason measured the elevation of the seas; ICESAT would investigate the role of melting ice in the rise.

Townsend had a lot to do with the Jason decision, including the naming of the satellite, which happened to be that of his son.[10] But the name also pertained to "Jason and the Argonauts" of Greek mythology. Zwally, who had initially cast his lot with EOS, won his plea for a specialized satellite for polar ice.[11] ICESAT stood for ice, cloud, and land elevation satellite. Its prime "goal was to use a laser altimeter to measure elevation changes on the ice sheets of Greenland and Antarctica."[12] These decisions took place over the 1994–1996 period. They were driven by science, money, Goldin's faster, better, cheaper emphasis, and the obvious significance of sea-level rise. Goldin helped protect the restructuring process from growing controversy surrounding climate change.

The Politics of Climate Change

While NASA restructured EOS, a wave of conservative Republicans seized control of the House of Representatives in January 1995. The new lawmakers strongly opposed government regulation. Their animus extended to environmental regulation and included the science that strengthened the case for regulating CO_2 emissions. As Goldin had warned Kennel, many legislators regarded climate change as bogus, scientists crying wolf to get money and wanted to cut back on the research that bolstered global warming as a problem. This was especially the case for NASA's MTPE, widely seen as Vice President Gore's pet program.

The republicans' point man for attacking MTPE was Robert Walker of Pennsylvania, chairman of NASA's authorization committee. One day Walker called Kennel into his office. He told him that whenever something appeared in the media about controversial issues like ozone depletion or climate change, he got "people in my office that want me to do things that I don't want to do, and they're telling me stuff and I don't know what to say to them." While Walker was indirect, Kennel got the point. He interpreted the discussion as an admonition to be careful "not to go out on a limb on any of these issues." It was OK to "get information out," but not to "go out making a big deal about it, just publish it."[13] That became his de facto science communication policy.

Walker was not satisfied by Kennel's unofficial policy of "scientific self-restraint." In April 1995 he asked the National Research Council to take a hard look at MTPE and the soundness of its scientific program. When NRC subsequently gave MTPE high grades, Walker remained on the offence. He attacked the program budgetarily.

Goldin strongly defended MTPE, as did Kennel. Walker continued his criticism until he left Congress at the end of 1996. Kennel departed NASA at the same time.[14] Although his budget had come down, Kennel avoided the worst cuts. He had restructured EOS and saved most of its science. Moreover, NASA made decisions to develop satellites critical to sea-level rise. Sea-level rise now had a focus within an overall program. However, it remained contested.

In December 1996, NASA and CNES signed a memo agreeing to collaborate on Jason. They said it was "imperative" to continue. "Data continuity" was essential because it not only would prove the *reality* of sea-level rise, but would give evidence of the *rate* of sea-level change. EOS had had as a goal a "predictive capability" and that was still a long-term goal for sea-level rise, as well as other aspects of an Earth System Science. Many in NASA and CNES referred to Jason as Jason-1, a fact signifying potential continuity of a program. However, to assuage Goldin, who insisted on change, the roles NASA and CNES played in TOPEX/Poseidon would be reversed. NASA would launch and CNES would build the satellite. Thus, there would be continuity *and* change.

Consolidation

After the tumultuous changes from 1992 to 1996, there was a desire on NASA's part for some stability in the program. Bill Townsend was the steady hand that could provide that, and he served as Acting Associate Administrator for MTPE for two years, 1996–1998. Within the limits of his "acting" role, Townsend consolidated the shifts in program organization that had transpired and changes to which he had contributed, such as those involving sea-level rise.

Under Tilford, NASA's MTPE had represented a concentrated model of satellites focusing on EOS. Under Kennel, it had shifted to a mixed approach, with three multi-instrument satellites (called Terra, Aqua, and Aura), along with a number of smaller satellites for specific issues, such as sea-level. Aqua, the water satellite, could detect the warming ocean. As the ocean warmed, it expanded. While it had other uses, Aqua could thus contribute to the overall sea-level mission. Many observers still called the total system EOS, while others saw only the three larger satellites as EOS and the remainder as independent. There was ambiguity, but there had been real change from a concentrated to distributed approach in the direction of intermediate and smaller entities.

Extending TOPEX/Poseidon

One of the issues that came up during Townsend's time as Acting Associate Administrator dealt with TOPEX/Poseidon extension. There had already been one extension, from three to five years under Kennel with Townsend then giving the go-ahead. This issue was a second extension and was much more controversial. In 1997, TOPEX/Poseidon was clearly still operating well. While the costs of extension were modest, such a decision went against the grain of Goldin policy which favored putting money into new ventures. Hence, extension was not automatic, even though Townsend favored it.

One of the strongest advocates for extension was Wunsch. With Tilford gone, he decided to return to his promotional role and made a strong case for extension. He ran into a challenge from a budget-conscious NASA official who contrasted spectacular missions under the planetary science program with lesser views of planet Earth. "Where is your volcano?" this official asked. He referred to a widely publicized, striking photograph of a volcano erupting on Io, a moon of Jupiter. It had made the front pages of newspapers, capturing considerable public attention.[15]

Wunsch had lots of scientifically exciting data, but no single picture that would convey to non-scientists what made the oceans so gripping to individuals like himself. Then, luck intervened. TOPEX/Poseidon spied a large El Nino developing, causing a surge of ocean crossing the tropical Pacific. El Nino was a major weather and climate maker. It could cause disasters across the world—too much rain in some places, not enough in others. An El Nino some years before had caused $2.2 billion in damage, in part because of a lack of warning and preparation.

TOPEX/Poseidon provided a warning now, and NASA advocates of extension prepared dramatic animations. Patzert, back at JPL, explained the El Nino pictures to the media and public. Becoming a regular television interviewee, he proved an outstanding science communicator. He also made use of the internet revolution underway to make the case for TOPEX/Poseidon's criticality. Wunsch and fellow advocates had what they needed—their version of a "volcano."[16] Wunsch later wrote that El Nino and NASA's response to it, "saved the extension."[17]

GRACE

Most that Townsend did that worked in favor of understanding sea-level rise was behind the scenes. A decision that was openly his was support for a pair of satellites called GRACE. In May 1997, Townsend backed this project (Gravity Recovery and Climate Experiment) which was a joint effort proposed by non-NASA researchers in the United States and Germany.[18] It used gravity to determine "mass," eventually including the shift of mass from glaciers and ice sheets to the seas. This use was one of several and not the most obvious at first. But in time, this application made it an extremely important complement to ICESAT.

Lindstrom

The position that Wilson had held at the time he left NASA in 1992, that of program scientist for oceanography, was finally filled under Townsend with a "permanent" official. Eric Lindstrom, who assumed the position in 1997, was a self-described "sea-going oceanographer."[19] He had been with the now-ended WOCE and then NOAA and came to NASA at the suggestion of Wilson. He would stay until 2019, providing a continuing effort behind the sea-level rise mission.

Defending NASA

Throughout Townsend's period as Acting Associate Administrator, climate politics festered. One issue involved a problem with TOPEX/Poseidon. In late 1995, Nerem, now at the University of Texas at Austin, had published an article in the *Journal of Geophysical Research* that reported on new TOPEX/Poseidon data.[20] It showed sea-level rising faster than earlier stated. There had been a sudden surge. Nerem subsequently raised some questions about the satellite data. NASA investigated and found a software error on the satellite which it fixed.

However, this meant that Nerem had to make a correction publicly, and he did so. The problem for NASA (and Nerem) was the way the media reported the mistake. In the July 1996 *New York Times* was the headline: "Error Inflated Estimate of Rising Sea Level, Researchers Report."[21] An August 1996 headline in *Science News* read: "Reining in Estimates of Sea-Level Rise."[22]

"Unfortunately," Nerem stated in retrospect, "that error gave ammunition to climate skeptics."[23] Townsend had to defend the accuracy of TOPEX/Poseidon and NASA's scientific objectivity.[24] The incident reminded NASA of the high media and public interest in sea-level rise and importance of getting the science "right." Climate politics was a constant factor in the NASA environment, no matter who was in charge.

From Townsend to Asrar

Goldin wanted an Earth scientist to head MTPE and hence Townsend moved to the deputy position at Goddard with a major task in managing the engineering development of ICESAT and other satellites. Goldin chose as the new Associate Administrator for Mission to Planet Earth Ghassem Asrar. Asrar had been chief scientist for the original EOS and had participated in its redesign.

When Townsend left in 1998, it seemed that MTPE had gotten through its most difficult times and was ready to move forward with an experienced scientist leader.

Notes

1. Richard Leshner and Thor Hogan, *The View from Space: NASA's Evolving Struggle to Understand the Home Planet* (Lawrence, Kansas: University of Kansas Press, 2019), 150.
2. Shelby Tilford, interview by Rebecca Wright, Oral History, EOS Collection, NASA, June 23, 2009. Retrieved from tinyurl.com/m6z22kz2
3. Charles Kennel, interview by Sandra Johnson, Oral History, EOS Collection, NASA, Oct. 21, 2002.
4. Charles Kennel, interview by author, Oct. 16, 2021.
5. Jeff Goodell, *The Water Will Come* (NY: Little, Brown, and Co., 2017), 62.
6. Steve Nerem, interview by author, Nov. 5, 2021.
7. Malcom Browne, "Most Precise Gauge Yet Points to Global Warming," *The New York Times* (Dec. 20, 1994), C4.
8. Ibid.
9. Charles Kennel, interview by author, Oct. 6, 2021.
10. Bill Townsend, interview by author, Dec. 8, 2020.
11. Correspondence, Bill Townsend, Aug. 19, 2020; See also Jay Zwally, interview by author, Sept. 28, 2021.
12. Jon Gertner, *The Ice at the Edge of the World* (NY: Random House, 2019), 243.
13. Charles Kennel, Oral History, EOS Collection, NASA, Oct. 21, 2002.
14. Richard Leshner and Thor Hogan, *The View from Space* (Lawrence, Kansas: University of Kansas Press, 2019), 178.
15. Carl Wunsch, "TOPEX/Poseidon as a Science Mission," in *History of Meteorology, Atmosphere, and Ocean Science from Space in France and Europe by its Actors* Ed. by Fellous, Jean-Lewis., (Paris, France: Institut Francais d'Histoire de l'Espace, forthcoming).
16. William Patzert, interview by author, Nov. 12, 2020.
17. Carl Wunsch, "TOPEX/Poseidon as a Science Mission," in Fellous; See also Carl Wunsch, interview by author, Sept. 13, 2020.
18. Byron Tapley, interview by author, Nov. 2, 2020.; See also Byron Tapley, interview by Rebecca Wright, Oral History, EOS Collection, NASA, January 12, 2010.
19. Eric Lindstrom, interview by author, Aug. 28, 2020.
20. R.S. Nerem, "Measuring Global Mean Sea-Level Variations Using TOPEX/Poseidon Altimeter Data." *Journal of Geophysical Research* (Dec. 15, 1995).
21. "Error Inflated Estimate of Rising Sea Level, Researchers Report," *The New York Times* (July 30, 1996).
22. "Reining in Estimates of Sea-Level Rise," *Science News* (Aug. 17, 1996).
23. Steven Nerem, interview by author, Nov. 5, 2021.
24. "TOPEX/Poseidon Sea-Level Measurements to be Revised," NASA News Release (July 25, 1996).

CHAPTER 6

High Hopes

Abstract This chapter covers the 1998–2001 period. Ghassem Asrar became associate administrator and changed the name of his division from "Mission to Planet Earth" to "Earth Science." He wished to emphasize the science and fly under the intensifying political radar of climate change. He built on TOPEX/Poseidon and Jason-1 success by getting a decision for a Jason-2 follow-on. This would feature a new technology called wide-swath altimetry. Goldin was adamant that NASA do R&D, not operations, or routine monitoring. He wanted innovation, firsts. Asrar also wanted to emphasize "the new" and sought to transfer Jason-2 to NOAA and its European counterpart, the European Organisation for the Exploitation of Meteorological Satellites (EUMETSAT) when it was mature. Meanwhile, he continued the NASA–CNES alliance. All seemed well.

Keywords Ghassem Asrar • Earth Sciences • Jason-2 • Wide-swath altimetry • R&D operations • The European Organisation for the Exploitation of Meteorological Satellites (EUMETSAT)

In February 1998, Ghassem Asrar became MTPE Associate Administrator, as Townsend moved to Deputy Director of Goddard. Asrar, with a PhD in Environmental Physics, had as his prime task to continue implementing the major changes made in the wake of EOS rescoping, including those

W. H. Lambright, *NASA and the Politics of Climate Research*, Palgrave Studies in the History of Science and Technology, https://doi.org/10.1007/978-3-031-40363-7_6

involving sea-level. It soon became clear that even more change was in store. At Asrar's suggestion, Mission to Planet Earth was renamed Earth Science Enterprise.[1] The name Mission to Planet Earth "was too political, too much Al Gore," he said. He wanted to "depoliticize" the program.[2] Asrar was very much an advocate for building what he called a "science of the Earth." He strongly favored emphasizing that science drove NASA's Earth mission, not politics. But the politics of climate would not go away, thanks in part to Gore's efforts at Kyoto in 1997 to get a climate treaty and congressional opposition. As climate policy came up in prominence, climate science became more controversial and sea-level issues along with it.

Asrar guided Earth Science from 1998 to 2006. In that period, he got the three major EOS satellites up, Jason-1 launched, and Jason-2 on its way. In line with Goldin's policy, he worked to transition ocean satellites from R&D to operations, starting with Jason-2. He expected operating agencies—NOAA in America and EUMETSAT, its counterpart in Europe—to pay at least part of the bill. The question of who did what became more complicated, however, in view of major decisions by Goldin and NOAA Administrator James Baker that would shift much leadership in climate observation from NASA to NOAA.

Asrar's tenure was sharply divided. From 1998 to 2001, he served under Goldin. From 2001 to 2006, he worked under republican appointees Sean O'Keefe and Michael Griffin. He went from a time of high hopes for Earth science in one period to one of great frustration later.

Relinquishing EOS Follow-Ons

Goldin's policy was clear, and Asrar adhered to it. NASA had a Mission to Planet Earth, but it was an R&D mission. That policy had significance for all NASA Earth Science programs. As Goldin saw it,

> in order for him to go after the next question of technological breakthrough in Earth science instruments, he needed to free up funds. But, unfortunately, the science community was increasingly viewing NASA as responsible for providing long-term operations with the science instruments. Hence, this meant NASA was becoming increasingly an 'operational' agency responsible for maintaining legacy systems. Goldin didn't want that.[3]

At the same time, climate change was rising as a policy issue, with Gore a powerful promoter and NASA constituent. Goldin was supportive of working on climate-research satellites, but guarded against continual operations. That policy came into play in terms of EOS and the Jasons.

In 1998, Goldin moved ahead with a grand bargain with James Baker, Clinton's NOAA Administrator. This was the same well-connected Washington-based oceanographer who had helped Wilson get his ocean satellite program started. As head of NOAA, Baker was ambitious to see his agency, with a fraction of NASA's budget and embedded in the Commerce Department, play a bigger role in oceans and climate.

Goldin wanted NOAA to take on climate and ocean satellite "operations." Thus, he willingly relinquished the proposed second and third sets of EOS satellites that were still tenuously on the NASA agenda following the Kennel–Townsend period. Baker was eager to assume responsibility for these satellites, or at least their instruments. Goldin and Baker agreed on a new satellite policy, and the White House and Congress concurred.

Berrien Moore, a University of New Hampshire Earth Scientist who was extremely active in EOS rescoping, climate change policy, and NASA-NOAA affairs, recalled the Goldin–Baker deal as a pivotal event. Baker, he related, saw climate as a major issue rising on the nation's agenda and believed NOAA should become not only a weather, but a climate agency. As Moore noted:

> Jim [Baker] recognized that it would cost money, and I think he felt the Clinton Administration and especially Al Gore would help him get the money. If Baker could get started in building a climate mission, money would follow.[4]

The Goldin–Baker approach applied to sea-level satellites as well as others related to the atmosphere. NASA, meanwhile, via the Jet Propulsion Laboratory, was exploring the next generation of ocean altimetry, the one called "wide swath ocean altimetry." This technology would look far more broadly and precisely at ocean–coastal interactions, the area of most important impacts. Asrar wanted to develop this new technology, work with CNES to insert it onto a Jason-2, and bring NOAA and its European counterpart, EUMETSAT, into a four-agency partnership. As NASA developed the next-generation altimetry, NOAA and EUMETSAT would eagerly pull it into operations, or so the strategy went.

Goldin told Congress that NASA could measure sea-level rise but needed a long-term research program to distinguish natural sea-level change from that which was a result of human-caused climate change. Both NASA and NOAA would move forward in climate change and sea-level rise, and they would share the mission and costs with NASA doing R&D and NOAA operations.

Not everyone agreed with this policy. Moore did not. Nor did Lindstrom. They believed the distinction between R&D and operations was overdrawn. Lindstrom did not like the term "operations" as applied to climate change or sea-level rise. He called it "sustained research." In slow-moving ocean matters, he believed, research had to be long-term and even repetitive. It had to be sustained. One could learn from monitoring and seeing trends. Goldin seemed to understand that fact, but he insisted NASA was about "firsts" not what he called "cookie-cutter" satellites. There was also more to this decision about roles. Goldin wanted additional money for NASA's robotic Mars program. With Goldin, the planetary program had his enthusiasm. Its director was an excellent salesman and Asrar was not his equal in this respect. But what really mattered was Goldin's personal passion for Mars.

NPOESS

In any event, Goldin was Administrator and had shown he did not care for resistance to his decisions. Baker was equally eager to move forward in remaking and elevating his agency. Prior to his being named NOAA's director, he had pushed with Sally Ride to extract NOAA from the Commerce Department, an organizational location that seldom worked to NOAA's advantage. That strategy had failed. But he and Gore had engineered a huge NOAA-DOD program to develop and deploy together the next operational weather satellite system called NPOESS (for National Polar-orbiting Operational Environmental Satellite System). It would merge separate civilian and military weather satellites. When Goldin cancelled the second and third sets of EOS satellites, their climate instruments were supposed to migrate from NASA to NPOESS. EOS would be succeeded by NPOESS. Townsend believed it imperative that EOS climate sensors not be lost in the Goldin–Baker arrangement. He worked with the White House Office of Science and Technology Policy to smooth the technological transitions.

NASA became a junior partner in NPOESS, helping NPOESS to be a weather *and* climate system that would save money and do more. It was up to Asrar and Lindstrom to implement the collective vision of Goldin and Baker and help make NPOESS work. At the same time, they had to implement Goldin policy on the Jasons. By adding wide-swath altimetry, they had Goldin's support for an "innovative" Jason-2 that would eventually transfer to NOAA and possibly NPOESS.

Enlisting EUMETSAT

In this context of larger governmental organization changes, Asrar acted to further his program. While having discussions with NOAA, particularly about ocean satellite policy, Asrar also opened a dialogue with EUMETSAT. In 1998, not long after his taking over what was now called the Earth Science Enterprise, he met with Tillman Mohr, the director of EUMETSAT, who was in Washington. Asrar asked—"Would EUMETSAT, as an operational agency, be interested in joining with NASA and CNES in a Jason-2 altimetry project?" Asrar said NOAA would probably also be a partner. Mohr wasted no time in responding: "I would have great interest," he declared.[5] He noted it would take a while—EUMETSAT was a pan-European agency representing many nations—to work out matters but Mohr was definitely positive. Like Baker of NOAA, he knew his agency had to change with the times, and that meant adding climate missions.

Moving Toward the Poles

Asrar also cared very much for developing ICESAT and GRACE. He wanted to begin shifting more of his overall research program toward the poles. This was where global warming was having its most immediate impact. He instituted an aircraft monitoring program that could periodically monitor arctic events to get a "baseline" of information prior to satellite work.[6] He adapted to the larger NASA-NOAA environment, seeing the NPOESS focus good in theory, but problematic in practice. The biggest issue was that much that was begun and planned for the future—both for NASA and NOAA—depended on climate change becoming a major national priority with more money and strong White House leadership. While there was much uncertainty, there was also room for optimism, including the sorting out of institutional roles. A good deal was underway.

In addition to what NASA was doing, NOAA's Baker had directed Wilson, who had earlier organized sea-level research while at NASA, to lead NOAA in building a multi-national system of high-tech buoys in oceans around the world. These would complement satellite observation in detecting warming seas. Satellites only measured ocean surfaces. The buoys could get at below-surface temperatures. Wilson called the system ARGO. There were plans for it to team with the Jasons, as in the Greek tale of Jason and the argonauts. Hopes were high for a coherent inter-agency and national policy for climate change, as the presidential election in November 2000 took place. Agency leaders believed they had moved partnership forward at their policy level. They needed larger policy to allow them to take the next step. Assuming Gore won, NASA and NOAA expected significantly enhanced support for climate science.

Notes

1. Ghassem Asrar, interview by author, Dec. 14, 2020.
2. Ibid.
3. Richard Leshner and Thor Hogan, *The View from Space: NASA's Evolving Struggle to Understand Our Home Planet* (Lawrence, Kansas: University of Kansas Press, 2019), 168.
4. Berrien Moore, interview by Rebecca Wright, Oral History, EOS Collection, NASA, Apr. 4, 2011.
5. Tillman Mohr, "EUMETSAT's Contribution to Jason-2—Its First Optional Programme," in *History of Meteorology, Atmosphere, and Ocean Science from Space in France and Europe by its Actors* Ed. by Fellous, Jean-Lewis., (Paris, France: Institut Francais d'Histoire de l'Espace, forthcoming).
6. Curt Suplee, "Melting Accounts for 7% of World's Sea-Level Rise," *The Washington Post* (July 21, 2000), A03. Retrieved from https://icesat.gsfc.nasa.gov/icesat/press_release/greenland.html

CHAPTER 7

"At Risk of Collapse"

Abstract In the 2001–2006 period Asrar's situation deteriorated. While Asrar got sea-level rise satellites up, along with the three larger satellites that had emerged from the EOS downsizing, he had serious funding problems assuring future missions. Also, Goldin sought to shift EOS climate follow-ons to NOAA and its huge weather satellite system, NPOESS (National Polar-orbiting Operational Environmental Satellite System). The Jason-2 wide-swath technology failed to develop. James Hansen sounded the alarm on climate change and criticized scientists for "reticence," especially about sea-level rise. When NASA Administrator Sean O'Keefe reorganized Earth Sciences under a Science Directorate, Asrar's science constituency saw his influence falling. O'Keefe's successor, Michael (Mike) Griffin, seemed to question whether NASA should even do climate change work. Outside NASA, scientists revolted and got a National Academy of Sciences National Research Council (NRC) Decadal Survey underway lest the Earth Science program "collapse" in their word.

Keywords The National Polar-orbiting Operational Environmental Satellite System (NPOESS) • James Hansen • Sean O'Keefe • Michael (Mike) Griffin • National Academy of Sciences National Research Council (NRC) Survey

W. H. Lambright, *NASA and the Politics of Climate Research*, Palgrave Studies in the History of Science and Technology, https://doi.org/10.1007/978-3-031-40363-7_7

On January 20, 2001, George W. Bush, not Al Gore, took command of the White House following an election so close it had to be decided by the Supreme Court. The coming of Bush, rather than Gore, did not mean the end of ambition on the climate and sea-level front. But it did put a damper on the highest hopes of NASA and NOAA.

The Bush Administration was willing to support research on climate change, but not give such research special priority. Some in the administration—and particularly in Congress—did not care for the research at all. For scientists inside and outside government, there was a wariness about the new political environment. Asrar carried on with the strategy he already had as best he could.

On November 2001, Goldin left NASA after a record-setting nine-year tenure. With his intense personality, sharp intellect, and distinctive management philosophy, he had made a difference, including for Earth Science. In December, Bush's appointee as NASA Administrator, Sean O'Keefe, took his place.

O'Keefe was a moderate Republican in a conservative administration. His major concern prior to February 1, 2003 was the International Space Station (ISS) and its huge overruns. From February 1 until he left NASA in early 2005, O'Keefe's agenda was largely defined by the Columbia shuttle disaster that took the lives of seven astronauts and scattered debris across a number of states. O'Keefe guided NASA's response to the accident. He also used the disaster to persuade President Bush to proclaim a "Vision for Space Exploration" in January 2004 that would move human space flight out of low Earth orbit, back to the Moon, and on to Mars. This Bush decision was macro-policy for NASA, causing almost everything else to take secondary—or worse—place in funding.

In this environment, Asrar at first did well—he and O'Keefe got along, and when O'Keefe came in, he spoke of NASA's need to preserve "the home planet." But as time went on, and O'Keefe became preoccupied with Columbia, Asrar struggled. When O'Keefe gave way to Michael Griffin as NASA Administrator in 2005, Asrar's situation deteriorated. His program's scientific constituency became alarmed, especially when the dominant NASA-NOAA-NPOESS climate strategy now in motion fell apart.

As 2006 dawned, Asrar was gone leaving the Earth Science program in crisis.

Initially: Good News

In December 2001, the news for NASA, and Asrar, was good. Jason-1 launched from Vandenberg Air Force Base in California. It went into the same orbit as TOPEX/Poseidon, which was still flying. It flew just slightly behind, ready to take over when the older satellite succumbed to age.

At the same time, NASA made it known that there would be long-term continuity with a new satellite. It would be formally named Ocean Surface Technology Mission (OSTM) but was universally called Jason-2. It would involve NASA, CNES, EUMETSAT, and NOAA. Exactly how the partnership would work remained to be seen. But the framework was clear, a testament to Asrar's diplomatic work.

In line with his strategy, the research agencies (NASA and CNES) would bring Jason-2 to a point where the operational agencies (NOAA and EUMETSAT) could take over. It was a "win-win" situation, or so it seemed. *Space News*, a leading trade journal in space policy, celebrated the partnership in an editorial headlined "Finally, Some Good News." *Space News* wrote approvingly: "This is the way the system is supposed to work."[1]

As for Jason 1, Asrar was exultant: "This particular mission is key to a major goal of the NASA Earth Science Program—that is, trying to understand how Earth's climate is changing and what are the consequences of these changes on human beings. That is the best gift that we can provide generations to come to [help] them establish sound economic and policy decisions when it comes to dealing with environmental issues." Lindstrom joined, saying: "This view of the oceans allows scientists to place shipborne measurements like never before. Present satellite altimetry as represented by TOPEX/Poseidon and Jason-1 is the lifeblood of modern oceanography."[2]

Innovating Jason-2

Getting Jason-1 up was a joy for NASA. There was also good news with the three EOS satellites on which Asrar had worked prior to being Associate Administrator. He had gotten land-oriented Terra up in 1992. He launched Aqua, a multi-instrument satellite for water, in 2002. Aura, with instruments related to the atmosphere, would launch in 2004. But he did not rest with these. He cared about the smaller, specialized satellites as well. The Jason-1 mission was projected to last for five years. He and his associates had to get another new start decision to prevent an interruption in the long record scientists said they absolutely had to have to credibly decipher the dynamics of sea-level rise.

Moreover, if all went well, Jason-2 would introduce a new technology: wide-swath ocean altimetry. TOPEX/Poseidon and Jason-1 deployed a narrow beam. As the name implied, wide-swath altimetry would have a broad-brush approach and sweep over seas and coasts. Since coasts were where the stakes were highest from sea-level rise, the significance of this new technology was clear, and NASA even spoke of pioneering a new satellite-based "coastal oceanography."

NASA had JPL working hard on the technology, and there was much enthusiasm for it. However, it was experimental and needed considerable money and time. The Earth Science enterprise did not have enough of either. The highest priority was data continuity. Asrar reached out to NOAA to help with funding and got nowhere. NASA also asked DOD to use a surplus launch vehicle it had in storage to save money. DOD declined.

Synergy for Sea-Level Rise

While worrying about Jason-2 developments, two key new polar-oriented satellites relevant to sea-level rise were ready for launch. GRACE went up in 2002 and ICESAT in 2003. These embodied Asrar's emphasis on the poles. Greenland especially was seeing huge sheets of ice collapse and fall into the ocean. GRACE was able to measure the transfer of ice-sheet mass to ocean water. ICESAT could show the loss of ice from views above.

These new satellites provided the synergy NASA needed to better understand what was happening and why. It was integrative knowledge. Sea-level rise was a result not only of warming water but the melting of ice, and GRACE and ICESAT could measure the glacial contribution to sea-level rise. Along with NOAA's ARGO system, still being developed, and newer tide gauges, there was coming into being a vast technological system for surveilling sea-level rise. In fact, what NASA was finding as a result of these satellites (such as TOPEX/Poseidon, Jason-1, GRACE, and ICESAT) was causing some in the agency to be extremely alarmed about the sea-level dimension of climate change.

The U.N. had established an Intergovernmental Panel on Climate Change, an international assemblage of scientists, to assess the nature and threat of climate change. It was an authoritative body to report to policymakers periodically on what was happening and its implications. It was not a body to perform research but to explain what that research portended.

Some in NASA thought that the IPCC was not taking NASA's sea-level research seriously enough. There was debate within NASA about how to communicate science findings. The informal policy, established by Kennel, was to be absolutely sure of the facts, present them in peer-reviewed journals, and publicly, but not get into policy advocacy. Asrar subscribed strongly to this view.

But a leading NASA scientist under Asrar believed the science was clear and disturbing enough to sound an alarm. This was James Hansen.

James Hansen and Asrar

Hansen had spoken to Congress and the media about climate change in 1988, and then, with the Clinton/Gore Administration in power, been relatively quiet. With Bush in office, and aware of the information coming from NASA's satellites, he decided he needed to speak up much more forcefully, even accusing other scientists of "reticence" in the face of facts, especially when it came to sea-level rise. He eventually wrote a paper, entitled "Scientific Reticence and Sea-Level Rise."[3]

Jay Zwally, a colleague and friend of Hansen, the prime scientist behind ICESAT, admitted to being himself unwilling to go the Hansen route, which attracted both publicity and criticism. Like most scientists, Zwally did not want to be called "alarmist." He was also cautioned by supervisors at Goddard, where he worked, to be careful or "you'll have people following you around, like Hansen."[4]

Hansen was not about to keep quiet about his concerns, and his outspoken approach to scientific communication went against Asrar's desire to keep his program under the political radar.

In July 2003, Hansen was asked to give a presentation to O'Keefe. This would be the first Earth Science presentation the NASA Administrator had received on the subject of climate change. Hansen had become increasingly alarmed about sea-level rise due to the melting of polar ice. He did not believe IPCC in particular was taking it seriously enough in its climate assessments. NASA was indeed taking research seriously, but not sounding a communications alarm. Hansen began to do exactly the latter in his papers and talks. He was working on a paper entitled: "Can We Diffuse the Global Warming Time Bomb?" when he briefed O'Keefe on the climate crisis he saw looming. As he later wrote of his presentation to the NASA Administrator:

> I sent a copy of 'Can We Diffuse the Global Warming Time Bomb?' to Ghassem Asrar, NASA's Associate Administrator for Earth Science, who would accompany O'Keefe on his visit to Goddard [where the presentation was to be made]. As we were assembling in the Goddard Director's office for this presentation, Dr. Asrar showed me a version of the paper that he had given O'Keefe—Asrar had changed the title to something about 'climate change' to make it less 'incendiary.'

Hansen's main conclusion from his paper was that "the stability of the Antarctic and Greenland ice sheets would surely set a low limit on permissible global warming and thus a low limit on greenhouse gasses." The paper and presentation used the word "dangerous" in describing the threat the Earth faced. O'Keefe's only comment during the presentation was that Hansen not use the word "dangerous" because there was uncertainty about climate, and no one knew what constituted danger. That was all O'Keefe said, and Hansen did not take it as an order. He wrote that it came across as "unequivocable advice."[5]

Asrar knew climate change was a serious problem. But he felt Hansen was going too far to get attention for his views. Hansen "did not understand the difference between science and politics," in Asrar's opinion.[6] If Asrar worked to tone down Hansen's rhetoric, however, he failed. NASA's Public Affairs Chief, Glenn Mahone, took over the effort and was equally unsuccessful. The Hansen issue continued under O'Keefe's successor and became an ongoing headache for Asrar.

Earth Scientists Complain

If Asrar had reason to be frustrated with Hansen, it was partly due to the fact the controversy his rhetoric engendered worked against Asrar's aim to grow his program's budget. Also, in the wake of Columbia, and its aftermath, Earth Science was impacted along with other NASA programs. In January 2004, President Bush announced his "Vision for Space Exploration." It was back to the Moon, on to Mars and beyond. That became the agency's dominant priority.

As a result of a potentially transformational new mission, with inadequate resources, O'Keefe in 2004 sought economies through reorganization. One consequence was putting planetary and Earth science under one overarching science directorate. That was the original model Goldin had changed in favor of more independent status for Earth Science. The return to the earlier approach had the effect of making Asrar a Deputy Associate Administrator for Science rather than Associate Administrator for Earth Science. That looked like a demotion of Asrar and their field to many in the Earth science community. But Asrar did not complain, at least publicly.[7]

Nevertheless, there was little question that Earth science was perceived as falling on harder times. The Earth science community, looking at what was happening at NASA, saw a diminishing number of future satellite programs ahead. Where were the "new starts" they asked. Where was the strategy? Asrar was getting the first (and only) set of EOS satellites up at the turn of 2000. What of NASA's post-EOS future?

Leaders of the community pressed NASA and Congress and got the National Research Council to agree to produce a "Decadal Survey" listing the major Earth science missions the U.S. government should sponsor in the ensuing ten years. Other disciplines had used decadal surveys to their advantage. Earth science had to do this as well, they believed.

Partnering Problems

The NRC decadal survey was about space-based Earth observations in general. Ocean satellites were one component. In a sense, the latter were doing well, thanks to TOPEX/Poseidon and Jason-1. In another way, there were questions and the need for future new starts. Also, there was the Jason-2 problem. The plan for it to incorporate wide-swath technology was not working out. For various reasons—time, money, technology, all of the above—NASA decided in 2005 to build Jason-2 with the same technology as in Jason-1. That would save money to be sure and maintain data continuity, but it did not move technology forward, as NASA and scientists had hoped.

Perhaps the most frustrating part of the Jason-2 decision-making process had to do with the partnership aspect that looked so promising to many in 2001. CNES was anxious as before to work with NASA on Jason-2 development. EUMETSAT was still eager to take on Jason-2 when it was ready but was taking a long time to get a decision from its many-nation leadership. But the real problem as far as NASA, and particularly Asrar and Lindstrom were concerned, was NOAA.

Baker's successor as head of NOAA was Conrad Lautenbacher, a retired Admiral. Lautenbacher was as ambitious as Baker had been about making NOAA as critical to the nation in climate as it was in weather. He inherited the chosen vehicle for doing so, NPOESS. Thanks to Goldin and Baker, the climate instruments intended for the second and third sets of EOS satellites were now planned to go to NOAA for NPOESS. NPOESS, like the original EOS, was a huge set of satellites, big science, costing multibillions over time. Comprehensive, it was seen as taking over all operational ocean altimetry eventually. In theory, at least, it could solve the research-to-operations challenge, by allowing NOAA to assert a powerful "pull" for NASA R&D results.

But NPOESS was descending into severe delays and cost overruns. The two key partners, NOAA and DOD, were not getting along. The result was that NOAA was depending on money that it did not have and could not get. Baker had hoped that the money would come as climate got to be a national priority, under Gore. Under Bush, it was not a priority.

Asrar had planned to get NOAA to share costs for Jason-2. Instead, NOAA asked NASA for money.[8] Lautenbacher said he was an "ardent supporter" of the Jason series and called it a "golden opportunity" to enlarge NOAA capability and role.[9] NASA wanted to see words turn into a budget.

Michael Griffin as Administrator

At the beginning of 2005 O'Keefe had left NASA for the presidency of Louisiana State University. In April, Michael Griffin took his place at NASA's helm. Griffin, an engineer with multiple degrees including a PhD, saw his prime task designing the hardware that would take NASA back to the Moon and on to Mars, a program called Constellation.[10] He also had to return the shuttle to flight in the wake of Columbia and resume construction of the International Space Station. With human space flight dominating his attention, he did not have much time for other programs, including Earth Science. He came across to many as not particularly concerned about climate change and its impacts, such as sea-level rise, or NASA's role in these endeavors. To increase money for human space flight, he cut science programs.

Decadal Survey Interim Report

As Griffin took power at NASA, the NRC published its Decadal Survey Interim Report.[11] Its message was ominous: the satellite infrastructure that the nation had constructed for Earth observation was "at risk of collapse." Existing satellites were getting older and had to be replaced. New measurements were needed. Jason-2 was a hopeful sign, but required a new start decision, as did other satellites. Decisions could not be postponed. The co-chair of the NRC panel, Berrien Moore, testified to Congress, painting a dire picture and also emphasizing potential "collapse."[12]

On May 30, 2005, *Space News* reported that the optimistic picture it had painted of interagency cooperation for Jason-2 in 2005 had not materialized. What was supposed to be a "showcase example" of a "research-to-users success story" was not in evidence.[13] Consequently, with the satellite future so hazy, the goal of "data continuity" that advocates so desperately needed to track and predict sea-level rise might fail to be reached.

Asrar's view was that his Earth science community and others had to be patient. The projects earlier planned would come, but they would take time, given budget constraints.[14] The community was not willing to wait. The president's Moon/Mars decision and Griffin's Constellation priority seemed to be a threat to science, in general, and Earth science, in particular. The community saw Asrar as lacking influence within NASA.

Asrar was indeed in a weakened position under Griffin and Bush. One of his colleagues likened him to a "governor of a state in an occupied territory."[15] The reference was to a widespread perception of Bush Administration hostility to climate change and thus research related to it. Offhand remarks by Griffin and an increase of NASA Public Affairs Office pressure on Hansen contributed to this sense. Asrar decided to leave NASA as 2006 dawned for a presumably less stressful position at the Department of Agriculture. Lindstrom and others carried on, while their NASA superiors looked for a successor to run Earth Science. It would not be easy to find one.

TOPEX/Poseidon Ends

Given the quest of ocean scientists for data continuity, and thus new starts, it had been fortuitous that TOPEX/Poseidon had lasted so long, and Jason-1 seemed equally robust. But time inevitably took its toll. A milestone for ocean observation occurred on January 5, 2006. On that day, TOPEX/Poseidon came to an end after more than 13 years of activity. It lost its ability to maneuver and stay in orbit. It had created a revolution in oceanography and a breakthrough in documenting the reality of sea-level rise and climate change.

It had also shown the value of inter-agency collaboration across the Atlantic. It had stimulated advocates in NASA and their allies, including those in France, to push fervently to continue satellite observation of the seas. TOPEX/Poseidon had attracted thousands of users in science, the military and industry through its images. Ocean scientists had written a multitude of papers advancing their field.

With Jason-1 operating, the data continuity users wanted was assured, at least for a while. However, there was worry about Jason-2. Advocates inside and outside NASA needed a decision to move it forward. While they worked toward that goal, the Hansen controversy came to a head.

Hansen Controversy Climaxes

Hansen had not let up on his attempt to communicate the seriousness of climate change to a wider audience than fellow scientists. NASA's Public Affairs Office, now headed by David Mould, tried to edit what he wrote and screen media interviews and presentations. Hansen resisted all controls and had his own lines of contact with the media.

On January 20, 2006, *The New York Times* carried a front-page story headlined "Climate Expert Says NASA Tried to Silence Him." That story drew immediate rebuke to NASA from Sherwood Boehlert, the republican congressman who chaired NASA's House authorization committee. He told Griffin to cease any scientific censorship. Griffin responded February 3 with a formal statement directing Public Affairs not to "alter, filter, or adjust engineering or scientific material produced by NASA's technical staff." He fired George Deutsch, a 24-year-old political appointee and Mould assistant who had sought to muzzle Hansen.

Hansen believed Deutsch was directed by higher-ups—Mould and even Griffin. But Boehlert did not want to press the matter.[16] Although Hansen continued to make his case, the controversy surrounding him diminished. It pointed up, however, the fact that NASA's climate research, including that provided by ocean satellites, had high media and public interest. The issue was how to communicate risks the science discovered. NASA looked for a middle ground between scientific reticence and scientific alarms.

Adopting Jason-2

To the relief of many, NASA finally forged a decision to adopt Jason-2. On April 10, 2006, NASA, NOAA, CNES, and EUMETSAT signed an agreement to cooperate on development, launch, and operation of the satellite. In asking for funds from the White House and Congress, NASA stressed Jason-2 was a "bridge" project. The four agencies would share costs.

NASA promised to adhere to its R&D agenda and let the operating agencies take over funding for a potential Jason-3 completely. NOAA and EUMETSAT seemed willing to call Jason-2 a "bridge" so long as NASA and CNES paid the bulk of the bill. NASA and CNES did pay most of the $300 million for building Jason-2 and the operating agencies lesser amounts. NOAA paid the least. Lindstrom, exasperated with NOAA, declared that it "did not have the money or political will."[17]

The NPOESS Crisis

The need for continuity was painfully clear, although NOAA's role was not. NOAA's problems with Jason-2 stemmed in large part from its misery involving NPOESS. That crisis reverberated to everything else it did and to NASA and European partners as well. In 2002, when NPOESS managers awarded a prime contact to Northrup Grumman, the NPOESS cost had been projected at $6.8 billion for six satellites. In June 2006, the cost was up to $11.4 billion, and number of likely satellites down to four.[18] NPOESS managers, under fire from the White House and Congress, decided in that month to remove the climate instruments. These were to come from NASA after the second and third sets of EOS satellites were cancelled. NPOESS was envisioned an operational EOS in addition to weather, a multi-purpose set of satellites. The climate instruments were expected to include an ocean altimeter for operational observations. This would not happen now.

Ocean observation and sea-level rise measurement remained exclusively part of the Jason series, now extended to Jason-2. The failure of NPOESS to fulfill its ambitions kept ocean satellites on their separate trajectory, under NASA, with an unclear operational destination. A Jason 3 could keep measurements going, but its adoption was by no means a certainty.

Reactions

A congressional hearing investigating NPOESS followed. It revealed the disappointment and anger many scientists felt about the program. Richard Anthes, President of the University Corporation for Atmospheric Research and Co-chair with Moore of the NRC Decadal Survey Panel, testified that the panel had envisioned NPOESS as the long-term operational foundation for much of Earth observation. "Obviously, it's not the foundation we thought it would be."

NOAA chief Lautenbacher admitted that "difficult choices and tradeoffs" had to be made. The climate instruments were "secondary sensors." He expressed hope they might go back on if money became available. Speaking for NASA, Griffin said the threat for climate science was not immediate. The problem would come in 2010 or 2011 when existing equipment phased out. What NASA would do given NPOESS's contractions required a plan for the future that did not exist. "Right now...we don't have a plan," Griffin stated. NASA was waiting on the NRC report, but that any new missions needing to develop necessary hardware "would require money not in the budget."

Anthes said the NPOESS restructuring "throws a monkey wrench" in NRC's own process of producing a report since the panel had not anticipated the "huge loss." He declared that NRC's interim report had said "that our Earth observing system was at risk of collapse, and that is before we knew what was happening with NPOESS. The collapse is now happening before our eyes. We can only hope that it is the bottom and over the next few years we will start climbing back out."

Not everyone lamented the NPOESS decision to drop climate. Hansen said it was a mistake to try to insert this kind of research into NPOESS. "The attempt to put climate measurements into the weather satellites was a terrible idea right from the beginning—a bureaucratic solution that was doomed to fail eventually."

Michael King, an EOS project manager at Goddard, argued that climate researchers had not lost everything due to NPOESS. "One example is [the study] of sea-level variability with radar altimetry." He pointed out: "We've got a data record going back to 1991 or 1992 of global sea-level and its variability and rise via TOPEX/Poseidon and Jason-1." In fact, he went on, the deletion of the altimeter from NPOESS was "not a bad thing because it was the wrong orbit for an altimeter anyway."[19]

Where Next?

With NOAA's removal of climate sensors from NPOESS, this meant return of many of them to NASA, with an unexpected funding burden. Also, NASA faced another dilemma as did the whole Earth observation community. NASA saw its role as producing first-of-its-kind satellite technology that, when proven, could transfer to operating organizations for extended use. Jason-2 was not first of its kind, but third of its kind, harkening back to TOPEX/Poseidon. While this sequence kept sea-level measurements going, it was to some extent a holding action. Jason-2 was supposed to be a "bridge." But a bridge to where? The NPOESS destination seemed to have been removed as an option.

What was to be done when it came to NASA and climate awaited guidance from the NRC panel, Griffin said. What NASA would do with that guidance depended in large part on who would be the next leader of Earth science at the space agency. But who would want the job?

Notes

1. "Finally Some Good News," *SpaceNews* (Dec. 17, 2001).
2. "NASA to Launch Jason 1, Timed Earth Science Satellites," *Aerospace Daily*, Nov. 21, 2001.
3. James Hansen, "Scientific Reticence and Sea-Level Rise," *Environmental Research Letters* (May, 2007).
4. Jay Zwally, interview by author, Sept. 28, 2021.
5. James Hansen, *Storms of My Grandchildren* (NY: Bloomsbury, 2009), 77–79.
6. Ghassem Asrar, interview by author, Dec. 14, 2020.
7. Tony Reichhardt, "Earth Science Loses Autonomy as NASA Switches Focus to the Moon," *Nature,* (July 1, 2004), 430, 3. Retrieved from https://doi.org/10.1038/430003a
8. Ghassem Asrar, interview by author, Dec. 14, 2020.
9. Conrad Lautenbacher, Oral History, EOS Collection, NASA: *History of Meteorology, Atmosphere, and Ocean Science from Space in France and Europe by its Actors* Ed. Fellous, Jean-Lewis., (Paris, France: Institut Francais d'Histoire de l'Espace, forthcoming).
10. W. Henry Lambright, *Launching a New Mission: Michael Griffin and NASA's Return to the Moon* (Washington, DC: BM, 2009).
11. National Research Council, *Earth Science and Applications from Space: Urgent Needs and Opportunities to Serve the Nation*, Interim Report, (Washington, DC: NAS, 2005).

12. Berrien Moore, interview by Rebecca Wright, Oral History, EOS Collection, NASA, Apr. 4, 2011.
13. Peter B. de Selding, “Data Continuity May Be Affected By Delays in Jason-2 Development,” *Space News* (May 30, 2005), 1.
14. Ghassem Asrar, interview by author, Dec. 14, 2020.
15. Mark Bowen, *Censoring Science: Inside the Political Attack on Dr. James Hansen and the Truth of Global Warming* (NY: Dutton, 2008), 128.
16. Ibid.
17. Eric Lindstrom, interview by author, Aug. 28, 2020.
18. Jeremy Singer, “NPOESS Restructuring Plan Trims Satellite Capabilities,” *Space News* (June 12, 2006), 6.
19. Jeremy Singer and Brian Berger, “NPOESS Loses Research Luster,” *Space News* (June 12, 2006), 6.

CHAPTER 8

Rebuilding Begins

Abstract This chapter covers the period 2006–2009. Michael Freilich took over Earth Science. An oceanographer, he regarded Wilson as a mentor. He used the Decadal Survey (2007–2017) as a guide to planning and strategy for insulating his program from disruptive climate politics. He began incrementally to rebuild the Earth Science division, pointing to the sea-level program as a priority. Promoting new satellites, he made sure there was continuity in polar ice observation through an ICESAT-2 and an aircraft program, ICEBridge, between ICESAT-1 and this follow-on.

Keywords Michael Freilich • ICESAT-2 • ICEBridge

In early 2006, Jack Kaye, the head of research for NASA's Earth Science Division, called Michael Freilich, Associate Dean, and professor in the College of Oceanic and Atmospheric Sciences at Oregon State University. He asked: "Do you know anybody who wants to come to [NASA] Headquarters to run the Earth Science Division?" Freilich, 52, had two grants under the division and had been relatively close to NASA his whole career. He had gotten his PhD in oceanography from the Scripps Institute of Oceanography, University of California, San Diego. He spent the initial years of his career at NASA's Jet Propulsion Laboratory (JPL) and was an early recruit to the new field of satellite-based ocean science. In 1992, at

W. H. Lambright, *NASA and the Politics of Climate Research*, Palgrave Studies in the History of Science and Technology,
https://doi.org/10.1007/978-3-031-40363-7_8

the suggestion of Stanley Wilson, whom he considered a mentor, he joined Oregon State, which was establishing one of the first programs in the country in satellite oceanography. An active researcher as an academic scientist, he had recently moved into university administration. He had been involved in assisting NASA in the EOS restructuring period and knew NASA officials and leaders in the Earth science community in the United States and abroad.

It was obvious to Freilich that Kaye was hoping he was interested in the job. Freilich knew about the problems Earth science had at NASA and had a comfortable niche at Oregon State. He and his wife had two grown children who lived in the East. The idea of being closer was appealing. But the real reason, he told Berrien Moore, was this: "Earth science is important to the future of the planet. It is important to my family. The job is important to my sense of calling. I can't back off."[1] He felt a personal mission to give direction to a program that needed leadership. He called Kaye and told him: "Yeah, I'll come to Headquarters."[2] He took a leave of absence from Oregon State, gave up his NASA grants, and agreed to a two-year stint in Washington. Arriving November, he stayed from 2006 to 2019.

Instilling Confidence

The Earth science community, including the Earth Science Division of NASA, was demoralized. He had to convince everybody he could turn the program around. As a researcher, he had gotten to know many Europeans. They trusted him. Taking the program's reigns in November, he quickly secured the NASA–CNES alliance. Fully aware of how important those ties were to TOPEX/Poseidon and the Jasons, he understood that the French needed reassurance that Earth Science at NASA had a future given the disruptions involving Bush's Moon/Mars program and NPOESS divestment of climate plans. NASA needed the French connection, and CNES needed NASA for its own institutional security.

Given budget constraints, he expected "partnerships" to be critical for NASA's getting a program accomplished in the future, and the French partnership for ocean observation and sea-level rise was the model for the program as a whole. He worked with Yannick d'Escatha, the head of CNES, to engineer a new agreement between NASA and CNES. Promulgated in January 2007, it was broad and applied to NASA–CNES collaboration generally. It also helped him establish a relationship with Griffin, who signed the agreement on NASA's behalf. Freilich was a scientist and manager, but observers also noticed he seemed, instinctively, to have political skills. Relying on Kaye for day-to-day internal management, he clearly savored his outside role. A good communicator, he was "crisp, to the point, clear."[3]

Using the Decadal Survey

Freilich realized that scientists, NASA, and everyone else involved in Earth science needed a new sense of forward motion and commitment. NASA's Earth Science budget, which had once been $2 billion when the first Bush backed its MTPE, was down to $1.4 billion when Freilich took over. The Decadal Survey was a potential tool for showing NASA's various constituencies that there was a plan for the future.

Prior to coming to NASA, Freilich had been worried that the Survey could backfire. He regarded the Earth Sciences as having various scientific communities that talked past one another. He believed, as his predecessors did, that space satellites could help pull them together. But he worried that the Decadal Survey would show Congress and the administration they could not agree on priorities. But once in office, and after conversations with Moore and Anthes, he changed his mind. They were building a consensus among the communities. The Survey results could be, for him, not only a scientific "plan," but "strategy" in dealing with his political environment. The Survey added credibility to his own conversations with Griffin, the White House, and Congress.[4]

As a well-connected oceanographer, he knew the priorities in his field, especially in respect to sea-level rise and glacial melting. But his domain encompassed the atmosphere, solar radiation, land, water, and much more. Believing in the potential of an Earth System Science, the message he conveyed was that he wanted to construct an "integrated system" of satellites that could provide the best science for the money available. Initially, however, there was friction between Griffin and the two chairs of the Decadal Survey he did not need.

The NRC Decadal Survey

At the beginning of 2007, the NRC final report became public.[5] It echoed many of the alarms its 2005 interim report had sounded and repeated the fear of "collapse" of satellite observation unless the government acted to develop a new suite of satellites to succeed those ending their useful lives. It listed 15 satellite projects needed in the next decade. In an audacious move, the report assumed a 30% gain in budget to make up buying power that had been lost since 2000.

The panel charged that just at the point that the nation had realized capabilities to make major contributions to climate prediction on time scales of seasons to decades and to monitor changes in the ocean's health, it was in danger of losing many observations because of programmatic failures or lack of will to sustain measurements. It predicted that between 2006 and the end of the decade, the number of operating sensors and instruments on NASA spacecraft, most of which were well past their normal lifetimes, would decrease by some 40%.

Aside from money, the NRC panel cited the structural problem of agency missions. It was "particularly concerned about the lack of clear agency responsibility for sustained research programs and transitioning of proof-of-concept measurements into sustained measurement systems." It saw one problem being NASA's lack of mission requirements for extending measurements in climate-relevant fields.

To revitalize Earth observation, NRC called for developing a number of new cutting-edge research missions NASA could launch. Of the fifteen proposed, a portion was relevant to sea-level rise and its causes. In particular, it called for a wide-swath ocean altimetry mission. Jason-2 was supposed originally to incorporate this technology but had not done so. JPL had subsequently found the ocean-oriented mission could be augmented to include hydrology, expanding it beyond the ocean to inland waters. The NRC panel agreed and recommended a new satellite called SWOT, for Surface Water and Ocean Topography. It recommended its development and launch in the 2013–2015 time frame and estimated the cost at $450 million.

The causes of sea-level rise lay with thermal heating of the ocean and glacial melting. While this heat could be monitored by existing satellites and NOAA ocean-based buoys, the glacial contribution was a great unknown. The panel asked: "Will there be catastrophic collapse of the major ice sheets, including those of Greenland and West Antarctica, and if so, how rapidly will this occur? What will be the time patterns of sea-level rise as a result?" To answer such questions, priority had to go to developing an ICESAT-II along with a GRACE-II.

There were other questions and other satellites to address them. The panel dealt with many aspects of the Earth sciences, not just climate change and sea-level rise. The whole system needed resilience. The report's view was that just as the need had become unprecedented, the nation's Earth observation satellite program was in disarray.

Reactions

Griffin did not take kindly to the report or media reactions. Moore met with Griffin to discuss the document. Griffin looked at him and, as Moore recalled, stated: "Well, you've just asked for a bunch of money, and we had come up with this decline in terms of Earth science funding at NASA and we'd created this envelope of what we would do if we got back." Moore responded, that "we simply observed the fact that we'd lost almost 30 percent in terms of real dollars between 2000 and 2007."

"So," Moore continued, "when we did this decadal and we put a persuasive case back on the table, in some ways the Administrator's response to that—which was negative—saying we couldn't afford this, we couldn't afford that, we couldn't do these things because he wanted to go off to the Moon—gave us a real debating arena in which we could put our ideas out there. *The Washington Post* and *The New York Times* covered it because we had a disagreement."[6]

On January 16, *The Washington Post* headlined the report: "Cutbacks Impede Climate Studies: U.S. Earth Programs in Peril, Panel Finds." It quoted Moore: "NASA's budget has taken a major hit at the same time that NOAA's program has fallen off the rails. This combination is very, very disturbing, and it's coming at the very time that we need the information most."[7] *The Washington Post* followed up this story with an editorial attacking Bush's priorities: "Martian Logic: Earth Science Pays the Price for Starry-Eyed Ambitions."[8]

A few days later, Griffin defended NASA's program and priorities in a 2007 Goddard Space Symposium address. He called the NRC report "a rather brazen recommendation that more money be allocated to Earth science." He termed it a "clear attempt to upset the traditional funding boundaries between and among the various science portfolios at NASA." He charged that the choice between human space flight and Earth science was "false."[9] Moore and Anthes reacted with a defense of their report in an op-ed in the trade journal, *Space News,* May 29.[10]

The debate escalated two days later when Griffin was interviewed on national radio and commented on climate change, declaring: "I have no doubt that a trend of global warming exists. I am not sure it is fair to say it is a problem we [NASA] must wrestle with."[11]

Freilich kept quiet. He didn't have to say anything. Many others strongly criticized Griffin, including Hansen, who exploded that the comment "indicates a complete ignorance of understanding the implications of climate change."[12] Griffin's statement came, ironically, when Bush himself was back-tracking on the climate issue and trying to show he cared about it in spite of his withdrawal from the Kyoto Accord Gore had negotiated.

Griffin soon realized he had erred. He publicly apologized to NASA for what he had said. He put out a statement in which he sought to explain he meant NASA should conduct research on climate change, but not get into policy solutions.[13]

The impact of the NRC report, the flap over it, and controversy over Griffin's remarks combined to make more money for Earth science an issue for NASA, the administration, and Congress. Freilich had no illusions that he would get an immediate 30% increase. But he could use the new urgency and visibility for Earth science to get a badly needed increase in budget. It had indeed been going down in recent years. Griffin was now in a position where he had to find more money for Earth science research. Freilich was a beneficiary of Griffin's predicament.

Research Strategy

Freilich told NRC that the 30% raise was unrealistic in the near term, but that he would seek to grow the Earth science budget incrementally. He now had a roadmap, and he called the Decadal Survey a "Bible" which he would use. "I can't prove this is the right answer," he explained, "but I will put a little money in all the missions recommended in 2007 to execute a broad program. If I don't have the money, I'll have to pick the right ones."[14] Sea-level rise was obviously one of "the right ones" he selected for priority, a de facto flagship for a broad, science-driven program.

His strategy included a change in program structure. Under Asrar, Earth science was skewed in the direction of the three EOS multi-instrument, billion-dollar satellites (Terra, Aqua, Aura). As he thought about new satellites, Freilich wanted smaller, more specialized entities covering the array of NRC recommendations. This smaller-satellite distributed system would take more effort from him for integration, but also gave him more flexibility in investments.[15] As a byproduct, it would help him build a broader scientific constituency. He told associates: "My strategy was to have missions that would launch every year… success incrementally." These satellites would be "smaller, more sellable, and I could give the Administration examples of success."[16]

He was equally conscious of societal needs and climate change. Sea-level rise and glacial melting were clearly societal issues. How much he emphasized them in statements depended on his reading of his political environment. He had great respect for Hansen, but did not want to go that route. His general strategy was to let the science speak for itself.

Particularly difficult for charting a research strategy was the problem mentioned by NRC, namely, the blurring of roles between research and operations. Like his predecessors, he did not want NASA's role to go too far into "operations," the repetitive orbiting of Earth with the same kind of satellites. He told his scientific community: "The more operations, the less money for research satellites."[17] He did not want NASA to do what he thought NOAA should do. He wanted to work with NOAA, but not get sucked into an operations vortex. He would have to be careful, and he was conscious that in rebuilding the Earth science program, it would be as important in what he chose not to do, as what he chose to do.

Congress

In dealing with Congress, he was completely aware of the partisan difference in political points of view about climate change and couched his remarks accordingly. With most republicans he downplayed comments about climate change and with most democrats he played them up. He was careful how he handled the issue, however. A particular congressional constituency was interested in sea-level rise, namely those who represented coastal states. It cut across political parties. Legislators who might not favor "climate change" could appreciate "sea-level rise." Freilich told his deputy, Sandra Cauffman, "You have to tailor what you say to the audience. Find out what is important to them."[18]

In March, while the conflict between Griffin and NRC raged, Freilich testified that sea-level rise satellites were "one of the triumphs" of the Earth Science program and one of his "very, very high priorities." He pointed out that Jason-1 was flying and, under his watch, "the very first of the new missions that are in the budget for launch—and it will launch in 2008"—was Jason-2.[19]

Of all the lawmakers who focused on sea-level rise, none was more influential than Bill Nelson (D. Florida), a key senior legislator on NASA's authorization committee. In a July hearing, Nelson expressed concern about the divestment of sea-level and other climate measurements from the NPOESS constellation. Freilich explained that NOAA had to prioritize weather, but that NASA would deal with climate needs and mentioned sea-level in particular. Mary Kicza, top satellite official for NOAA, was at the same hearing and underlined Freilich's point about NPOESS emphasis on weather. However, she pointed out that NOAA was still into climate. It was a partner in Jason-2 from a standpoint of ground activity. And it was working with EUMETSAT to determine roles with a Jason-3.

Freilich did not get into the Jason-3 matter at the hearing, but he was involved behind-the-scenes in discussions about NASA's role in a possible Jason-3. With NPOESS losing its altimetry, a Jason-3 was now essential for data continuity. He saw Jason-3 as operational and not something NASA should do, however. Instead, he wanted to move ahead with the wide- swath altimetry research satellite, SWOT. This was something the NRC report also wanted.

Freilich intended to revitalize his program by a sharp focus on science. He intended for NASA to lead in satellite-based climate research, but in the way he wanted. His program would be broad, inclusive, integrated, balanced, cutting-edge, with sea-level rise as a top priority. He cared about all Earth sciences, but sea-level rise was the element the general public could best see as a signal for climate change, and it was vitally important to powerful lawmakers, not only Democrats but also Republicans from Georgia, Mississippi, and Louisiana. They had seen storm surges, floods, and what rising waters could do.[20]

Prioritizing ICESAT-2

At the beginning of 2008, NASA announced it would substantially increase investments in the Earth Science program over the ensuing five years. It then had 14 satellites in orbit. These consisted of the larger EOS three and a number of smaller, specialized satellites, including those related directly or indirectly to sea-level rise. Many were aging and needed successors. Griffin told Congress in February that NASA would begin the rebuilding with two NRC-recommended new starts, one of which was ICESAT-2.[21]

In April, Senator Barbara Mikulski (D. Md), the most influential democrat on the Appropriations Committee responsible for NASA, declared that the Decadal Survey priorities guided her own decisions as well as those of NASA. She specifically pressed Griffin on the ICESAT program, which was managed by Goddard, in her state. Griffin assured her that NASA would fund Earth Science more adequately, and decadal priorities such as ICESAT-2. ICESAT-1 was having technical problems that could end its life soon, at a time when ice-melting at the poles was becoming increasingly worrisome.

Jason-2 Launches

While NASA looked ahead to new starts, it reached a milestone with Jason-2, now ready to launch. On June 10, it soared into space. The satellite cost $449 million. NASA called it a flight facilitating the transition from development to operations. The four agencies involved had shared the $449 million cost, with NASA and CNES reluctantly paying the most for a satellite essentially duplicating Jason-1. The two agencies also paid for research associated with the satellite.

The mission went well. Not long after the launch, Jason-2 went into orbit close behind Jason-1. Both satellites circled about 830 miles above Earth. They entered a dual calibration phase that lasted five to six weeks. They then carried out activities in tandem for as long as possible. Jason-1 still functioned well, but there was no predicting its duration. Jason-2 was projected to last for five years.

While NASA, in publicity about the mission, mentioned relevance to climate change, the emphasis was on other uses, such as marine industry and oil and gas users.[22] The aim was to appeal to republican lawmakers. One problem Freilich did not face was interference from Public Affairs or some other NASA office. NASA's Inspector General a few weeks earlier had issued a report verifying and condemning the pressures Hansen had experienced. Thanks to this report, Griffin's earlier policy memo, and Hansen's own actions, that lowered his profile, Freilich did not face the internal issues that had bedeviled Asrar.[23]

Decision-Making for Jason-3

As Freilich contemplated how to implement the Decadal Survey, he knew what he did not want to do, namely fund Jason-3. There had to be a Jason-3. Everybody agreed that sea-level rise needed constant monitoring. But Freilich insisted that NOAA take over repetitive flights. He and Lautenbacher met and agreed on a full transition in roles. It was a role Lautenbacher wanted for NOAA, and he pledged NOAA would pay approximately $118 million, roughly half the estimated costs.

NOAA looked to its counterpart, EUMETSAT, to take the operational lead in Europe. Like NOAA, EUMETSAT wanted this task, and—like NOAA—it had limited money to assume a new mission, climate, while stressed to meet weather forecasting responsibilities, some of which could be urgent. In the late summer and early fall, EUMETSAT sought help within Europe. CNES said it would provide some spare satellite hardware, but no additional money. Like NASA, it wanted to protect funds for research and development and its identity as an agency looking to the frontier of technology.

EUMETSAT, as a pan-European organization, discussed possibilities for funding with the European Union and European Space Agency (ESA). ESA had been established years before by European nations to allow them to engage in large-scale projects beyond the scope of national programs. Akin to NASA, it developed large rockets and participated in the Space Shuttle and International Space Station. The major Europe-wide Earth observation effort was the Global Monitoring for the Environment and Security (GMES) Program. It was largely operational and had not emphasized ocean observation.

Germany objected to ESA's helping to fund Jason-3. That was a U.S.–French program, Germany insisted. Europe should not pay for a program whose benefits went to France. France was ESA's leading funder. Germany was second. Once Germany spoke up, Italy and others joined in opposition to ESA's involvement.

France (i.e., CNES) was furious. If Germany was not going to help France, through ESA, then France was not going to help Germany with projects it wanted via ESA. That position threatened many projects much larger than Jason-3. CNES's director, Yannick d'Escatha, declared: "We are not going to watch this [Jason program] fall by the wayside. It is not my nature to want to block this or pound the table, but the user community has been clear that data continuity is critical to our Earth observation efforts."

Decision-making on Jason-3 ground to a halt. In October, Jean Jacques Dordain, ESA's Director, became heavily involved, stating he "fully understood the French view." It was not possible, he pointed out, for Europeans to "trumpet" the cause of "data continuity" in satellites "and at the same time threaten ocean altimetry" by an inability to fund Jason-3. But he also avowed he understood the view of others that ESA's funding of Jason-3 would set a precedent that would be a problem.[24]

In November, France and Germany resolved their disagreement and ESA was enabled to be a partner. The compromise was to tie Jason-3 to bigger decisions pending about Europe's enhanced role in climate change. ESA would "repurpose" a satellite it had planned for another ESA-EUMETSAT mission. The "repurposed" satellite would be Jason-4, called officially Jason CS. The CS stood for "continuing service," a phrase embodying operations. The Jason CS satellite would be built by Germany. Jason-3 (built by France) and Jason CS would be absorbed into GMES and then its proposed successor, Copernicus.

NOAA closely followed the European decision process affecting Jason-3. Lautenbacher made frequent telephone calls and traveled to Europe as necessary.[25] NASA tried to keep out of the Jason-3 politics. However, it was in close contact with CNES. Freilich wanted to work with CNES on SWOT. Meanwhile, everyone involved in planning for the future of Earth and sea-level rise observations awaited the next U.S. president, elected in November. This was Barack Obama.

Notes

1. Berrien Moore, interview by author, Apr. 14, 2022.
2. "Michael Freilich (1954–2020) Former Director, NASA Earth Science Division," Profile, Retrieved from https://solarsystem.nasa.gov/people/432/michael-freilich-1964-2020
3. Sandra Cauffman, interview by author, Jan. 20, 2022.
4. Michael Freilich, interview by author, Feb. 18, 2020.
5. *Earth Science and Applications from Space: National Imperatives for the Next Decade and Beyond Final Report*, (Washington, DC: National Academies Press, 2001). Retrieved from https://www.nap.edu/read/11820
6. Berrien Moore, Oral History, EOS Collection, NASA.

7. "Cutbacks Impede Climate Studies: U.S. Earth Programs in Peril, Panel Finds," *Washington Post*, (January 16, 2007). Retrieved from https://tinyurl.com/9pr3b8b3
8. "Martian Logic: Earth Science Pays the Price for Starry-Eyed Ambitions," *Washington Post*, (January 18, 2007). Retrieved from https://tinyurl.com/5fhz3vm5
9. "Michael Griffin Address, 2007 Goddard Space Symposium," Mar. 20, 2007. Retrieved from https://www.nasa.gov/pdf/171973main_mg_Goddard_20070316.pdf
10. "Berrien Moore and Richard Anthes, Op-Ed," *Space News*, (May 29, 2007). Retrieved from https://spacenews.com/oped-fundamental-misunderstandings/
11. David Kestenbaum, "NASA Chief Assailed for Climate Comments," NPR (June 1, 2007).
12. Clayton Sandell and Bill Blakemore, "Scientists Surprised by NASA Chief's comments," *ABCNews*, (Nov 2, 2007).
13. Alicia Chang, "NASA Chief Regrets Remarks on Global Warming," *NBCNews*, (June 5, 2007). Retrieved from https://www.nbcnews.com/id/wbna19058588
14. Art Charo, interview by author, Oct. 23, 2019.
15. Michael Freilich, interview by author, Feb. 18, 2020.
16. Ibid.
17. Ibid.
18. Sandra Cauffman, interview by author, Nov. 20, 2022.
19. Michael Freilich Testimony, "Hearing before the Subcommittee on Space, Aeronautics, and Related Sciences of the Com. On Commerce, Science, and Transportation," *U.S. Senate*, (March 7, 2007), 169. Retrieved from https://www.govinfo.gov/content/pkg/CHRG-110shrg78568/html/CHRG-110shrg78568.htm
20. Walleed Abdalati, interview by author, Feb. 22, 2022.
21. "Commerce, Justice, Science, and Related Agencies Appropriations for 2009." *GovInfo | U.S. Government Publishing Office*, (2008). Retrieved from https://www.govinfo.gov/content/pkg/CHRG-110hhrg42708/html/CHRG-110hhrg42708.htm
22. NASA, OSTM/Jason-2 Science Writer's Guide, NASA History Files.
23. Michael Freilich, interview by author, Feb. 18, 2020.
24. Peter B. de Selding, "France Ties Its ESA Funding to Support for Jason-3 Program," *Space News*, (Sept. 22, 2008), 1, 4.
25. Conrad Lautenbacher, "The Beginning of Jason," in *History of Meteorology, Atmosphere, and Ocean Science from Space in France and Europe by its Actors* Ed. Fellous, Jean-Lewis., (Paris, France: Institut Francais d'Histoire de l'Espace, forthcoming).

CHAPTER 9

Gaining Momentum

Abstract This chapter covers the period 2009–2014. With the Obama Administration, the Earth Science program became more "climate centric." Freilich crossed swords with Deputy Administrator Lori Garver who pushed him to go from measurement to mitigation. NASA established an interdisciplinary Sea-Level Change Team (N-SLCT) to enhance its communications to the public. The climate sensors came off NPOESS and back to NASA. Freilich and NOAA agreed that NOAA would fund an operational Jason-3, while Freilich sought to put his money into a second-generation wide-swath sea-level satellite he and the Decadal Survey wanted, SWOT (Surface Water and Ocean Topography satellite).

Keywords Lori Garver • NASA Sea-Level Change Team (N-SLCT) • A "climate centric" program • Surface Water and Ocean Topography satellite (SWOT) • Jason-3

On January 20, 2009, a new administration anxious to step up the pace of action on climate change came to power. Barack Obama recognized the climate threat but had more immediate priorities with the "Great Recession" and Iran and Afghanistan wars. He signaled his climate interest by appointing John Holdren, a Harvard physicist and climate change advocate, as his Science Advisor. Space policy was marginal on his agenda,

W. H. Lambright, *NASA and the Politics of Climate Research*, Palgrave Studies in the History of Science and Technology, https://doi.org/10.1007/978-3-031-40363-7_9

but it was clear that he wanted NASA to give greater priority to climate change. It was not until July, however, that NASA got a new administrator, Charles Bolden, a retired Marine Brigadier General and former astronaut. Bolden was supportive of Earth science, but his passion was human space exploration.

Freilich operated at the mid-level of policy—the administrative level. There had been some talk among his friends about whether he should try to raise his office's status back to an Associate Administrator rank. Earth Science was under an overall Science Associate Administrator. However, Freilich said he could function in his present role as a Director of the Earth Science Division. Under Bolden, he had considerable autonomy to continue on his chosen course, implementing the 2007 NRC Decadal Survey. His goal remained an integrated Earth science system. He was especially attentive to climate change and sea-level rise. But he wanted to do it his way. Under the Bush Administration, White House pressures favored a low profile on climate change. In the Obama political context, the pressures were to do more, visibly, and tilt science toward solutions. Freilich adapted, but kept his eye on his Bible, the Decadal Survey, and steered his own course in a polarized political environment.

Meanwhile, the decision in Europe to transfer leadership on Jason-3 from CNES to EUMETSAT and Europe-as-a-whole was slowly underway. In America, NOAA had the lead in Jason-3 policy. Jane Lubchenco, a marine biologist, was now in charge at NOAA. Mary Kicza continued to head satellite operations. Freilich pushed forward to rebuild his program, keeping science to the fore and knowing that he was accountable to Congress as well as the White House. His stance in this respect put him at sharp odds with NASA's Deputy Administrator, Lori Garver, who wanted him fired.

Polar Melting

When it came to sea-level rise, the most important policy imperative for Freilich in 2009 was to deal with ICESAT-1. Its technical troubles had worsened and it would likely end its service some time that year. Freilich had gotten the Bush Administration to authorize an ICESAT-2 to investigate changes at the poles, but its development would take at least five years. He decided to fill the gap between ICESAT-1 and ICESAT-2 with an aircraft program. Asrar had initiated such a program, but it had lapsed. Freilich intended this one to be especially robust.

Early in 2009, he appointed Tom Wagner, a glaciologist who had been at the National Science Foundation (NSF), to run a cryosphere science program at headquarters. Cryosphere (ice) had its technical locus at Goddard, and Goddard still was deeply involved in management and research. This appointment gave emphasis to the activity at NASA, given the heightened concern for glacial melting as a cause of sea-level rise. He came to NASA, Wagner said, to work on climate change.[1]

Wagner recalled that when he walked in the door at NASA's Earth Science Office at NASA Headquarters, Freilich told him what he wanted: "Create a new mission to go to the polar regions." Freilich followed this up by telling Jack Kaye, his research director, "Find the money to pay for it." This was a NASA priority, he exclaimed, and he wanted no "gap" in polar observations. Wagner was soon at work on what became known as "Operation Ice Bridge."[2]

By May, planning for the new program was moving forward and Wagner was explaining its aims to the media. As its name implied, it was a "bridge" between ICESAT-1 and ICESAT-2. While aircraft could do an adequate job, they were no substitute for satellites, he emphasized. The latter could function constantly and cover a larger scale of polar terrain.[3] In August, as anticipated, ICESAT-1 failed to function and aircraft took over. Concentrating on ICESAT and other matters, Freilich hoped he was free of Jason-3, except for ongoing research funding. Building and using the satellite, he continued to argue, was up to NOAA.

NOAA's Role

NOAA had pledged to provide the U.S. share of Jason-3 costs and sought to get the money from the administration and Congress in 2009. Both were reluctant to grant NOAA's request. NOAA was still beset with woes from NPOESS. The Obama Administration was taking a hard look at NPOESS, which seemed to be needing ever more money and taking ever longer to develop. NPOESS cast a dark shadow over NOAA and its capacity to relieve NASA of Jason-3.

NOAA nevertheless wanted to play a leadership role in climate change science. In testimony before somewhat skeptical lawmakers, it touted its ARGO program. Completed in 2004, ARGO was now a widespread system of instrumented buoys in oceans around the world. It complemented Jason satellites in terms of measuring ocean warming. Along with ice melting, this warming caused seas to expand. Whereas satellites measured surface temperatures, ARGO sensed temperatures below the surface for a more complete picture of the effect of climate change on the seas.[4]

ARGO, like the Jasons, entailed international alliances. There were now some 3000 of these relatively inexpensive instruments doing their work, an achievement for NOAA that it could highlight in arguing for the Jason operations role. It was a role NASA wished to cede to NOAA. But lawmakers from coastal states, like Bill Nelson, were worried. They felt the stakes were high, and NPOESS gave them reason for concern about NOAA capacity.

Wilson, architect of ARGO, now a senior scientist at NOAA, was taking a lead within NOAA pushing for Jason-3. He reached out to his former ally, Wunsch. Since helping to get TOPEX/Poseidon extended, the MIT scientist said he had moved to the sidelines of ocean satellite lobbying. He told Wilson he had other interests, and an operational satellite like Jason-3 was simply not "exciting" for a research scientist like himself. Wilson replied that he understood.[5]

Obama's Policy Change

On February 1, 2010, Obama released his proposed FY2011 budget for Federal agencies. The budget announced the decision to terminate NPOESS and end its NOAA–DOD partnership. In its place, on the civilian side, NOAA was to build a Joint Polar Satellite System (JPSS). It would cost billions, but presumably cost less than NPOESS. The budget also included funds for NOAA's Jason-3 contribution.

NASA was a big winner in the budget, at least as proposed. The budget authorized a cumulative \$10.3 billion for Earth Science over the FY 2011–2015 time frame. The objective was to return NASA Earth Science to the \$2 billion buying power it had in FY2000. This was an increase of 30% from recent levels. Obama asked NASA for a report on how it would focus on climate change with additional money.

While NASA worked on its report in early 2010, congressional appropriations committees considered funding requests for Jason-3 and NOAA's organizational responsibility. In April, Senator Mikulski asked NASA Administrator Bolden about NASA–NOAA collaboration to move research to operations. Bolden indicated progress and pointed out that a joint working group between NASA and NOAA was studying the transition issue. The NOAA FY2011 budget had money for Jason-3, which was the first major outcome of joint planning according to Bolden. The NASA budget gave the go-ahead to SWOT development, a Freilch priority.

Meanwhile, the congressional watchdog, General Accountability Office (GAO), published a report the same month casting doubt that there was much of an interagency strategy for transition, and the result would be future gaps in satellite coverage. It was addressing NASA–NOAA interagency relations in general, with respect to the research-to-operations question, not just Jason-3, or SWOT.

Garver vs. Freilich

The issue of how NASA would focus more on climate under Obama was not just a matter of report-writing for NASA's Deputy Administrator, Lori Garver. She regarded herself as a prime agent of Obama at NASA. She expected Freilich to be accountable to the White House—and herself as a political appointee. She was non-technical in education, but steeped in science policy, having run Goldin's policy unit previously. She was politically active and had worked to get Obama elected, and then headed his transition team for NASA. Many in the White House Office of Science and Technology Policy (OSTP) were close associates of her.

She felt NASA should do more than calculate the impacts of climate change, such as sea-levels. She wanted NASA to tackle solutions to the problems it discovered. In effect, she would broaden NASA's mission toward mitigation.[6] She wanted NASA to be proactive and visible in climate change on behalf of the president. When she had worked for Goldin, he had told her: "If you love to be loved, you will not do the right thing." She took that advice to heart as Deputy. Years later, after NASA, she told *The New York Times* that while NASA had provided important science to help understand global warming, it had not been involved deeply enough into the search for solutions. She wanted "the brilliant scientists of NASA to do more than just take measurements."[7]

She expressed her views directly and forcefully to Freilich. Freilich pointed out that NASA also reported to Congress, and many legislators disagreed with Obama on climate policy. Garver and Freilich clashed, not only on policy but in personality and style.

Freilich had his own strong views about how to direct Earth science, and he held to them. Speaking of his relationship with Garver, he told his deputy, Sandra Cauffman, "We are like oil and water."[8] After one meeting with Garver, Freilich complained to Chris Scolese, the Senior NASA

Associate Administrator and de facto general manager. Visibly upset, Freilich might well have resigned. Garver clearly wanted him fired. Scolese calmed him down.[9]

In April and May 2010, she sought to hire Moore for a Chief Scientist role. Moore had left the University of New Hampshire and was directing a "Think Tank" devoted to climate change. He was clearly the kind of politically savvy scientist-activist on climate change she wanted. He might have come to NASA had she not told him of her desire to dismiss Freilich. They were having wine together at a Holiday Inn across from the NASA Headquarters. She told him she wanted him to take on Earth Science and the "first thing you should do is get rid of Mike Freilich."[10]

Moore, however, thought highly of Freilich and the notion of his being an instrument for Freilich's departure did not go over well with him.[11] Indeed, Freilich was insisting that science, as expressed in the 2007 Decadal Survey, was his guide, not politics, whether Republican or Democrat. Moore had been a co-leader of the 2007 Decadal Survey. He chose to return to higher education and a major position at the University of Oklahoma. Aware of the president's desire, Freilich did bend somewhat in the Obama/Garver direction, but in his own science-first way.

A "Climate-Centric" Agenda

In June, NASA released its report, "Responding to the Challenge of Climate and Environmental Change."[12] It showed how NASA would restructure and enhance its program to give greater emphasis to climate. The report touched on virtually all measurement possibilities, such as changes in land, water, atmosphere, polar ice, sea-level, solar radiation, and more. The message was clear. With a $10.3 billion budget over five years, NASA could do much more, more quickly, on climate. The report was heavy in technical emphasis but did touch on policy, such as need for international partnerships and importance in moving swiftly from research to operations.

Sea-level rise and polar ice were given attention through the concept of "climate continuity missions." This meant NASA would take on more responsibility—operations by another name—in certain areas of climate change. The Decadal Survey had recommended a GRACE-2 that would take some time to develop given its technical complexity. NASA proposed in this report a simpler interim GRACE project that would be akin to what it was doing between ICESAT-1 and ICESAT-2—that is, a "bridge" that would keep measurements going in this area of climate priority. The "climate-centric" plan proposed new satellites to address climate change in all its aspects. These included new sea-level related missions, notably SWOT.

Freilich saw the climate-centric report as another Bible, complementing, not replacing the Decadal Survey, which he often carried with him in dealing with important officials in his environment. Garver did not believe Freilich was going far enough on climate change solutions. In the end, Bolden sided with Freilich over his Deputy, with whom he had a number of differences. NASA's job, he insisted, was "to generate information, explain what it means, sit at the table, answer questions such as how do you turn things around. It is not NASA's job to do mitigation." Bolden noted that Freilich was consistent in protecting his program from "politics either from the left or the right."[13]

Reality Check

The high hopes of early 2010 for steep budget gains and realization of Decadal Survey and "climate-centric" desires hit a reality check in 2011. The mid-term elections placed Republicans in charge of the House of Representatives. The Senate remained under Democratic control. Republicans took a hard line on budgets, and the result was partisan conflict and inability to pass a federal budget on time.

It took months into 2011 for NASA and NOAA to get satellite money they sought. Meanwhile, they lived on a continuing resolution that restricted how much and on what they could spend. When the White House and Congress finally enacted a budget in early May 2011, NOAA faced a $4.6 billion budget for FY2011 that was $140 million less than it received in FY 2010. This cut forced NOAA to delay the launch of Jason-3 by one year.

NOAA Assistant Administrator for Satellites and Information Services Mary Kicza said the launch would now be in 2014. Moreover, Kicza stated that the budget cut would also mean delays for the Joint Polar Satellite System's first satellite of 18 months beyond its target. She warned that it took six to eight years to develop a complex satellite and that the United States would be left with a gap in severe storm warnings. The difference in cost between the Joint Polar Satellite System and the Jasons was huge, but both were put at risk, and what happened with NOAA had implications for NASA.

NASA also suffered, but less. The White House withdrew two satellites from the new starts NASA wanted. ICESAT-2, however, was protected. NASA's problem was that ICESAT-2 was going to cost more than twice the $300 million the Decadal Survey projected. Freilich scrambled to find $20 million of Obama's economic stimulus money NASA received to help get ICESAT-2 underway.

Freilich declared that NASA had gotten "almost everything we requested" in 2011. "What's left in the program is rather robust." Looking ahead, he stated that the agency and scientists had to be more pragmatic. "I think the key to get out of the fix in which we landed ourselves is to be relentlessly objective and realistic about what our budget prospects really are," he said. "We haven't spent nearly enough time developing the consensus that says, 'If we only have this amount of resources, here is what we should do.'"[14]

Looking for Partners

Freilich was compelled to look further for partnerships in the United States and abroad as a way to extend his budget. In 2012, he thought he had a new relationship with the Air Force that would help his budget and help him with his sea-level rise/glacial melting priority.

Freilich wanted to launch ICESAT-2 in 2016. The Air Force intended to launch a weather satellite then, and he and the Air Force worked out what Freilich called "a real sweet deal" by which he could co-launch ICESAT-2 for a relatively modest cost. But in April 2012, the Air Force backed out of the arrangement, saying it could not launch until 2020. Freilich said he needed ICESAT-2 sooner.

This issue highlighted the fact that not only satellites cost money but so did rockets on which they rose. When the "sweet deal" fizzled, Freilich had to seek an alternative, one he found, but for which NASA had to pay all on its own, costing much more than it would have with the Air Force.[15] This situation also showed the problem NASA had in fitting its priorities into budget reality. The NRC Decadal Survey had not only assumed a 30% raise in NASA spending but its estimates for the satellites it recommended were well under the real costs.

Meanwhile, scientists were using ICESAT-1 and GRACE data to determine the severity of glacial melting. The new findings made it all the more imperative for NASA to push its polar ice research forward. As it did so, it took some comfort in its not having to pay for Jason-3. In the fall of 2012,

NASA and CNES moved SWOT forward into early development, hoping to get enough funds to speed its creation.

In late October, 2012 Hurricane Sandy hit the northeast coast, ravaging New York City in its wake. This event did not seem, at the time, to be a major problem for NASA. But it mattered enormously to NOAA, responsible for the weather warnings, and eventually came to impact NASA.

IPCC

In 2013, the IPCC came out with its latest assessment of climate change.[16] It called global warming "unequivocal" and human causation equally likely. The 2013 report differed from previous documents in IPCC confidence. This confidence was due in part to the long record NASA now had of sea-level observations. IPCC had been somewhat hesitant to rely overly on such satellite data, wanting a longer record of trends. But it now felt the data were strong enough. The next year, it listed satellite observations among the information essential to its work.[17] NASA had wanted IPCC as a principal user and constituent and now had it.

Nelson Hearings

Another user of NASA satellite findings was Senator Nelson (D. Fla). Nelson was chair of NASA's Senate Authorization Committee and was especially attentive to the impact of sea-level rise on the state he represented, Florida.

On April 22, 2014, Nelson held an Earth Day hearing in Miami Beach to call attention to sea-level rise and its impacts on Florida.[18] Various officials from the South Florida region testified on the issues they faced. NASA provided expertise through Piers Sellers. Sellers was a PhD meteorologist and Deputy Director of the Science and Exploration Directorate at Goddard. He was also an ex-astronaut and favorite of Bolden.

Sellers went over what NASA was doing on sea-level rise and what it meant. When Sellers finished his testimony, Nelson exclaimed: "I think it is important to point out that Dr. Sellers' testimony is not modeling, is not a forecast, it is a measurement. What he has stated, in fact, has occurred. And so those who deny climate change and sea-level rise, here is the proof right here." Sea-level rise, he went on, was not something to worry about years from now, in 2100. "It is happening right now!"

Establishing a Sea-Level Change Team

For years, Lindstrom had an advisory group for his program that looked at ocean research generally. Wagner began talking with him about a separate group focused on sea-level rise. The two program officers had their own research dollars to spend and decided to establish a "NASA-Sea Level Change Team."[19] They envisioned it as a group of scientists from different disciplines (oceanography, glaciology, geology, etc.) relevant to the problem of sea-level rise. It could advise them and in turn communicate science and the threat to a broader audience. Its membership could change over time as the problem evolved, and research needs charged.

The first scientist to calculate the rise from TOPEX/Poseidon measurements, Steve Nerem, would be the chair of the group. The decision to have such group reflected a new emphasis in the Earth Science Division, not only on sea-level but also how to communicate it as a problem beyond NASA to regional and local officials. NASA was carefully adapting to pressures to do more than research and measure the sea-level problem. Wagner and Lindstrom agreed to provide a certain amount of their research money for this science communication effort to bolster it. Kaye, their superior, also contributed funds from his discretionary budget.

SWOT

While endorsing this extension of NASA's science-communication role, Freilich's priority lay with science. It was extremely important to him to get a decision to move SWOT to faster development. SWOT epitomized NASA's role to advance technology to answer pressing science problems. SWOT represented a new technology that brought together observation of all waters of the Earth, linked oceanography and hydrology, and ocean-coastal interactions. The cost, which CNES agreed to share, would be high, much more than the NRC estimate—well over a billion dollars. It was thus a sharp departure from Freilich's strategy to emphasize smaller, less expensive satellites, launched quickly. On January 6, 2015, NASA and CNES signed the documents necessary to commence full, joint development. After years of barely maintaining research on the new generation of technology which SWOT represented, Freilich told JPL to shift to a higher gear. The window of opportunity was at hand.

These events and decisions were indicative of Freilich's efforts to revitalize the Earth Science Division. They also indicated the division's broadening focus on sea-level rise and polar ice melting. Freilich continued to steer his program in a science-building direction and resist internal and external pressures to go too far beyond its research and development role, especially where he believed NOAA should lead. But events were making that disciplined stance more difficult to sustain.

Notes

1. Tom Wagner, interview by author, Oct. 16, 2021.
2. Ibid.
3. Debra Werner, "NASA Using Aircraft to Study Polar Ice While Awaiting ICESAT Replacement," *Space News*, (May 25, 2009).
4. W. Stanley Wilson, interview by author, Oct. 27, 2020.
5. Carl Wunsch, interview by author, Sept. 13, 2020.
6. Garver's view was expressed clearly in comments she made to *The New York Times* in 2021. See also John Schwartz, "Devoting His Skills To Benefit the Earth," *The New York Times* (Mar. 30, 2021), DI, 4.
7. Ibid.
8. Sandra Cauffman, interview by author, Jan. 20, 2022.
9. Chris Scolese, interview by author, Mar. 12, 2022.
10. Berrien Moore, interview by author, Apr. 14, 2022.
11. Ibid.
12. "Responding to the Challenge of Climate and Environmental Change," *NASA*, (Jun. 2010). Retrieved from https://pace.oceansciences.org/docs/climate_architecture_final.pdf
13. Charles Bolden, interview by author, Feb. 15, 2022.
14. Lauren Morollo, "Climate Satellite Programs Scarred in Budget Fight," *The New York Times*, (May 4, 2011).
15. Dan Leone, "U.S. Air Force Deferral of Last DMSP Upends Plans for Launching ICESAT-2, *Space News*, (Apr. 16, 2012).
16. Intergovernmental Panel on Climate Change, *Climate Change 2013: The Physical Science Basis*, (2013). Retrieved from https://www.ipcc.ch/site/assets/uploads/2018/03/WG1AR5_SummaryVolume_FINAL.pdf
17. Peter B. de Selding, "Unnecessary Timeout Pushes Jason-3 Launch to July or August." *Space News*, (June 15, 2007), 13.
18. "Leading the Way: Adapting to South Florida's Changing Coastline." Congressional Hearings, (Apr. 22, 2014). Retrieved from https://www.govinfo.gov/content/pkg/CHRG-113shrg94339/html/CHRG-113shrg94339.htm
19. Eric Lindstrom, interview by author, Aug. 28, 2020; See also Tom Wagner, interview by author, Oct. 16, 2021.

CHAPTER 10

Taking "The Lead"

Abstract From 2014 to 2016, the climate change/sea-level rise program was in crisis. Obama had killed NPOESS and replaced it with another giant weather satellite system, Joint Polar Satellite System (JPSS). In the wake of Hurricane Sandy (2012), Congress and the White House told NOAA to stick with weather and NASA to take charge of both R&D and operational climate satellites. NOAA and NASA resisted as Freilich feared losing research—NASA's priority—to operations. NOAA ultimately found money for Jason-3. But EUMETSAT and Europe generally were alarmed by the turmoil and delays afflicting Jason-3. The European Union (EU) had established Copernicus, a large weather and climate program, mainly operational. It wanted partnership with the United States on a Jason-4, called Sentinel 6 Jason CS. Seeing NOAA's lack of support, European agencies called on NASA to take the lead for the United States. Freilich adapted his strategy to what was a clear turning point and won support from Congress and the White House for a broader NASA role with Europe and Copernicus. The election of Donald Trump in November 2016 helped catalyze fast action on both sides of the Atlantic.

Keywords Joint Polar Satellite System (JPSS) • Hurricane Sandy • Congress • White House • Copernicus • Donald Trump

W. H. Lambright, *NASA and the Politics of Climate Research*, Palgrave Studies in the History of Science and Technology, https://doi.org/10.1007/978-3-031-40363-7_10

As Freilich looked for funds to fulfill the 2007 Decadal Survey menu of new satellites, to become more "climate centric," fund SWOT, and cope with cost-overruns and delays afflicting ICESAT-2, he continued to resist pressures from Congress and the White House to expand into operations. A lot of the internal and external pressure for extending NASA's role came because of exasperation with NOAA. NOAA seemed the logical operator and wanted the mission, but had continual money troubles and limited support from the Department of Commerce, in which it was situated organizationally. Freilich and Lautenbacher had agreed that NASA would not fund Jason-3 and NOAA would take the lead on behalf of the United States in Jason-3 partnership with Europe. Lautenbacher wanted to do what he said he would do. But doing that proved a great struggle. Freilich told Cauffman he wanted to spend his still-limited money on NASA's priorities, not NOAA's. But circumstances changed, and Freilich changed.

NOAA's Dilemma

By 2014 NOAA's Kicza was increasingly arguing that funding and other issues were creating the possibility of a gap in weather satellite activity between the existing system for tracking severe storms and the more advanced JPSS. Her arguments were taking crisis-alarm form, and the recent Hurricane Sandy experience was fresh in the minds not only of meteorologists but also media and politicians.

One who paid particular attention was Senator Barbara Mikulski (D. Md) and chair of the Senate Appropriations Subcommittee responsible for both NASA and NOAA. On June 5, 2014, the Appropriations Committee directed NOAA to give "priority" to JPSS and its "flagship weather satellites." This meant its climate work, specifically Jason-3, was not to take attention and resources from JPSS. But the committee said that NOAA's smaller missions, such as Jason-3, were also important, and should be supported. The solution, the committee said, was for NOAA to concentrate on its "core mission," namely weather, and let NASA take responsibility for Jason-3. That would mean both weather and climate satellite missions could advance.[1]

This was a big decision about roles, apparently made without much debate or visibility. Jason-3 was an operational satellite, essentially another copy of the original Jason, harkening back to TOPEX/Poseidon. This directive did not go well with NASA or NOAA. The bill did not become law owing to unrelated conflicts between the White House and Congress over budgets. However, the action sent a strong message from Congress NOAA and NASA had to take seriously. Mikulski in particular expected to be taken seriously.

The following year, 2015, the White House Office of Management and Budget (OMB) provided language in budget legislation to the same effect as had Congress, namely that NOAA should concentrate on weather, leaving climate satellites to NASA, with the implication that when it came to satellites, NASA had to deal with monitoring (i.e., operations) as well as research. This was a role NASA did not want, but it had it, starting with Jason-3, unless Congress and the administration changed their minds or Freilich could find a way to minimize what he clearly viewed as a threat to his research emphasis.

NOAA decided to fight the guidance and retain its role as lead agency at least for Jason-3.[2] It had expended considerable time and effort already in planning and negotiating with the Europeans and NASA. Its leaders fervently wanted to be part of the administrative apparatus to tackle climate change. The problem, as NASA stated privately of NOAA, was that "its aspirations exceeded its budgetary grasp." NASA's mantra was: "NASA does new things in new ways. We may do new things old ways, or old things in new ways. But never old things in old ways." NASA, and the career bureaucrats who worked for it, saw their agency as an engine for technological change and critical support for science, including climate science. The agency's science constituency reinforced this view.

Fighting

NOAA argued it could develop JPSS and also Jason-3. It, in fact, did ultimately resolve its JPSS worries and commence deployment. Jason-3, however, was more problematic. OMB and Congress did not want to provide the resources for Jason-3 NOAA needed. That fact led to more delay and frustration on both sides of the Atlantic.

Europe, meanwhile, had consummated the decision process set in motion by the dispute between France and Germany over Jason-3. Copernicus came into being in 2014. Under European Union authority, it constituted a program involving a number of organizations with varying stances on the research/operations debate. For Jason-3, the relevant operator was EUMETSAT. Anxious to proceed, it waited for NOAA to resolve its funding problem. Lautenbacher later wrote of his difficulties, noting at times he "despaired" that NOAA could keep the Jason series going. He had to make a number of transatlantic phone calls and trips.[3] He recalled various critical meetings to keep the project alive.[4]

NOAA was able to retain Jason-3 responsibility, but was not able to provide funding in the amount originally promised. To get enough to cover the launch, NOAA had to reallocate funds "from existing non-dedicated sources." In other words, it had to find the money from its limited discretionary account—since OMB and Congress were not about to help an agency whose mission they wished to restrict. The process caused schedule slippages and extra cost to European partners, and critical concerns, in particular on the EUMETSAT side.[5] Finally, on January 17, 2016, NASA launched Jason-3 and was reimbursed by NOAA. NASA and CNES provided funds for associated research.

While the Jason-3 experience did prove successful in keeping sea-level rise observations going, it was clearly not what anyone wanted to repeat for Jason-CS, scheduled to go under the Copernicus Sentinel series in 2020, with the European Space Agency (ESA) in the lead.

The Zwally Dispute

As NASA tried to sort out what it would do in the wake of the Congress–OMB–NOAA imbroglio, it found itself coping with more disputes arising from the politics of climate. In November 2015, NASA's Zwally had published an article in the *Journal of Glaciology* that said Western Antarctica was not losing ice, as many scientists believed, but actually gaining ice.[6] Immediately, other scientists challenged his work. He had used ICESAT-1 and critics called attention to limits of this technology. Scientists using GRACE, which measured mass rather than height, differed with Zwally.

Many scientists said that Antarctica as a whole was losing ice in view of the amount lost in the eastern part. Zwally held to his view, but made it clear that further global warming—which he expected—could change the balance between gain and loss.

What started as a scientific dispute and disagreement over what satellites could and could not do in glaciology became embroiled in the still-intense political debate about climate change. The article was used by climate change deniers. Fox News and other conservative media platforms celebrated Zwally as showing that climate change activists were exaggerating the threat of sea-level rise.[7] NASA was caught up in the scientific/political controversy, much as it had been when Hansen, who had retired in 2013, had made public statements. Hansen had deliberately wanted to influence the policy world, but Zwally had reported on what he considered a purely science finding. The problem was that whatever climate scientists said seemed to be used by one side or the other of the global warming debate, Bolden tried to deflect political conflict by saying that NASA did Earth Science, not climate change, a distinction Moore and others criticized as too great a stretch.[8]

Freilich Bends and Accepts "The Lead"

Freilich cared deeply about protecting NASA's role in pursuing cutting-edge science and technology. He had resisted, for years, the idea that NASA take over operations, routine monitoring, much less getting into solutions. It was not just a Freilich view. This was a NASA position, and it went back decades. It was a strategy to protect NASA and its scientific constituency. NASA Administrator Bolden, like predecessors, strongly supported this position. Pressured himself by the White House and Congress, Bolden declared: "We disagreed vehemently that NASA should do operations. This was not our job."[9] The scientific community understood and generally supported that view. More money for operations could mean less money for research.

But others, like Lindstrom and Moore, saw operations as a form of long-term research. The problem was that the IPCC and other bodies, including NASA's own Sea-Level Change Science team, were making it clear that climate change was extremely serious and needed long-term, continuous surveillance and policy action. The issue was who did what. Europe, but not America, seemed to have reached the point of establishing just such a long-term and more comprehensive program with Copernicus. But various European officials believed that the European commitment required a U.S. commitment. Partnership mattered symbolically, financially, and politically to the Europeans.

One of these European advocates for collaboration, Alain Ratier, the head of EUMETSAT, was deeply worried. He raised his concerns with Freilich on a number of occasions. He feared the United States might rely on NOAA for Jason-CS—which Europe called Sentinel 6 Jason-CS. He had watched NOAA struggle mightily to get a budget for Jason-3 and did not believe the U.S. –European partnership would survive unless NASA came through with leadership and money.

NOAA simply did not have the necessary political and administrative support for reliable partnership at least in this field. Individuals in NOAA, frustrated with their situation, admitted as much. Ratier told Freilich that if he did not "take the lead," he would be endangering the sea-level/climate change program on both sides of the Atlantic. Ratier was blunt and said: "Jason CS will not go forward with NOAA. You take it on!" Freilich got the point. He saw the big picture. He fully understood that R&D was not enough and that the U.S. sea-level science system was incomplete without long-term monitoring.

There was a significant gap in the U.S. administrative system, one Freilich believed hurt the United States and its role in climate policy. A "national policy" decision was desperately needed to fill it, one that clarified roles and provided resources to carry them out. The United States was not ready to make that overarching decision. He decided NASA could not hold back, and gave Ratier a "Yes!" He also warned that he needed funding to go with responsibility.[10] But once Freilich chose to go in this leadership direction, he moved directly and swiftly.

Using contacts he had nurtured within NASA and government generally, Freilich was able to obtain funds in the upcoming presidential budget for Jason-CS. "That probably saved the program," Ratier recalled, as it bolstered the case for funding Sentinel 6/Jason-CS in Europe. As much as some in Europe wanted it, others had different priorities. This project was competing with many needs in Copernicus.[11] This was thus a crucial decision about climate science priority and sea-level surveillance in particular for both the United States and Europe. It was also a precedent for a new, enhanced relationship, broader and possibly far more consequential than the long-standing NASA–CNES alliance.

Freilich drew on his political capital and reputation in Washington as a "straight shooter" who managed his program in an apolitical, nonpartisan manner. Once the money issue was clear enough with administrative leaders in OMB and elsewhere, Freilich engaged EUMETSAT, ESA, and NOAA in a four-party discussion on how to back Sentinel 6/Jason-CS. Freilich also got key leaders in the science community who worried about losing research funding aboard an advocacy coalition of support. Finally, he moved to win the backing of Congress, through staff and senior legislators who held sway over NASA's budget. Republicans were in charge of both houses. His argument to them was that "science was the only and undisputable goal of NASA."[12]

Trump as Catalyst

The prospects for Sentinel 6/Jason-CS looked favorable until November 2016, when Donald Trump shocked everyone by beating Hillary Clinton for the presidency. Trump was an avowed climate change denier and stated he intended to remove the United States from the Paris Climate Accords that Obama had negotiated. Would he kill the sea-level satellite program? The Europeans were worried—as were Freilich and others.

The bureaucratic processes necessary for the four agencies to move ahead with joint satellite development were complex and convoluted, exacerbated by the fact that each agency in Europe was pluralistic in leadership, and various nations had to agree on matters of policy and funding. In the United States, NASA had to get the State Department's formal blessing for an international project. Normally everything took time, lots of time, and advocates feared a President Trump or his appointees would stop the program in its tracks if they could.

Freilich and his allies, including NOAA, moved to get all the decisions in place before Trump took office. Ratier "convinced Jan Woerner, the then DG [Director General] of ESA and the Chairman of the EUMETSAT Council to accept that NASA proceeds immediately with the request for clearance to the State Department without prior formal deliberations by the ESA and EUMETSAT Council." Freilich pushed the State Department to agree to this accelerated decision process and it did. The Europeans expedited decision-making on their side.

On December 16, 2016, NASA, EUMETSAT, and NOAA signed a Memorandum of Understanding for partnership and ESA signed by correspondence soon after in January.[13] What they approved was a mission entailing two identical satellites to provide continuity of operational sea-level change observation through the 2020s. NASA and ESA would develop the technology and build the hardware. NASA would perform the launch. EUMETSAT and NOAA would operate ground stations and distribute science data. Decisions about funding would be worked out later.

There was no certainty Trump or his allies would identify Sentinel 6/Jason-CS "as a climate dedicated mission"—which it was—but the transatlantic advocates could not be sure. The election of Trump ended any ambivalence about NASA's taking leadership for the United States on the mission and potentially expanding into an operational as well as R&D role. Like it or not, NASA risked becoming not only a science and technology agency for climate but a broader climate operations organization.

Freilich led in taking that risk in adapting NASA's mission. It meant making compromises on the Sentinel satellite with ESA he might ordinarily have contested, because ESA wanted some of the more technically advanced aspects NASA savored. Fortunately, Europe was also willing to assume much of the operations role. Details in the new relationship would have to evolve. At this point of decision, speed was essential. Trump's election had triggered fast and coordinated action in the United States and abroad, with Freilich as the orchestrator.

Notes

1. "Making appropriations for Departments of Commerce and Justice, and Science, and Related Agencies for the fiscal year ending September 30, 2015, and for other purposes," (June 5, 2014). Retrieved from https://www.congress.gov/113/bills/s2437/BILLS-113s2437pcs.pdf
2. Stephen Volz, interview by author, Jun. 20, 2022.
3. Conrad Lautenbacher, "The Beginning of Jason," in *History of Meteorology, Atmosphere, and Ocean Science from Space in France and Europe by its Actors* Ed. Fellous, Jean-Lewis., (Paris, France: Institut Francais d'Histoire de l'Espace, forthcoming).
4. Ibid.
5. Alain Ratier, email to author, Apr. 5, 2022.
6. Jay Zwally, "Mass Gains of the Antarctic Ice Sheet Exceed Losses," *Journal of Glaciology*, vol. 31, Issue 230 (July 2015).
7. Shannon Hall, "What To Believe in Antarctica's Great Ice Debate," *Scientific American* (July 6, 2017).
8. Charles Bolden, interview by author, Feb. 15, 2022; See also Berrien Moore, Oral History, EOS Collection, NASA.
9. Charles Bolden, interview by author, Feb. 15, 2022.
10. Alain Ratier, interview by author, Mar. 20, 2022; See also Alain Ratier, email to author, Apr. 5, 2022.
11. Ibid.
12. Alain Ratier, communication to author, Feb. 23, 2022.
13. Alain Ratier, interview by author, Mar. 30, 2022; See also Alain Ratier, email to author, Apr. 5, 2022.

CHAPTER 11

Mounting Defense

Abstract The fear of many advocates was whether Trump would terminate the sea-level effort since it was clearly a climate mission. Freilich protected the sea-level satellites, including those for glacial melt observation, in behind-the-scenes negotiations with the Office of Management and Budget (OMB) and Congress. Trump did seek to kill a number of climate satellites each year he was president and failed as Congress put back the money he extracted. Freilich retired in 2019 and soon after, in 2020, died. Europe and NASA honored his memory by naming the satellite he had helped make possible Sentinel-6 Michael Freilich. It launched in 2020, with a duplicate to follow later in the decade. Institutionalization was achieved via a NASA–Europe alliance!

Keywords Office of Management and Budget (OMB) • Sentinel-6 Michael Freilich • A NASA–Europe alliance

On January 20, 2017, Donald Trump took office as President. His transition team included Robert Walker, who had headed the House Science Committee years before and been an ardent critic of the Earth Science Program. Walker remained a critic, especially of climate change research. That stance against climate change was profoundly seen in Trump, who called climate change a hoax and indicated he would not only get the

W. H. Lambright, *NASA and the Politics of Climate Research*, Palgrave Studies in the History of Science and Technology,
https://doi.org/10.1007/978-3-031-40363-7_11

United States out of the Paris Climate Change Accords but rescind EPA actions in this field. The political signals for NASA's Earth Science Program, and sea-level mission, were extremely unfavorable. Trump was in the White House, Republicans controlled both chambers of Congress. Scientists associated with the program were decidedly nervous and worried.

What was Freilich to do? He had slowly, painstakingly rebuilt the program since taking it over in 2006, then at a low point. He had an agreement with Europe to help pay for two launches of Sentinel 6/Jason CS satellites in the 2020s that would maintain a watch on sea-level rise. In fact, each satellite would cost one-half billion dollars and have improved sensors allowing closer surveillance of ocean–coastal interactions. He had follow-ons to ICESAT and GRACE soon to launch to calculate the impact of global warming on glaciers and ice sheets. And he had prioritized a new wide-swath technology for sea-level and inland water surveillance. SWOT was in development and would eventually to go into the enlarging and comprehensive space infrastructure for monitoring rising seas with even greater attention to their coastal impacts than the Sentinel 6/Jason CS satellites. There was a threat to the program and all he had done he would have somehow to avert.

Reassuring Scientists

On December 26, before Trump's inauguration, Freilich had an opportunity to speak to scientists attending an Earth Science town hall meeting at an American Geophysical Society Conference in San Francisco. His voice was broadcast from Cape Canaveral, Florida, where he was on hand for a satellite launch. Scientists filled every seat and lined the walls of the large room to listen, many fearing their grants and research careers in jeopardy.

Freilich stated that it was possible there would be cuts to the program. However, he pointed out, "It is not at all obvious that that is going to happen." He expressed optimism that "the benefits will be recognized as they have been in the past." He advised the scientists to do their work and let the data speak for them. "We don't do policy" at NASA, he declared, but "provide the information for all the policymakers to draw their own conclusions." He told them to avoid "spending time pining for what could have been."

Freilich noted that NASA's Earth Science budget historically waxed and waned depending on which party had political power. He reminded the audience that it had fallen from 11% of NASA's budget in 2001 to 6% in 2006. Only after the NRC called attention to the Earth observation system's "danger of collapsing" did the budget go up. The Earth Science budget had increased to $1.9 billion in 2016 from a low $1.4 billion in 2006. The program had been "revitalized." It had 19 major missions flying on satellites and instruments on the International Space Station. There were many more observations to come. He expected to launch "several missions per year."[1]

He promised to work with other agencies to augment funding. His immediate superior, Thomas Zurbuchen, head of the Science Division, who attended the meeting, also encouraged the scientists. He urged them to emphasize how useful their work was to disaster mitigation at the state level. Emphasize the practical, he said. Freilich and his associates would repeat again and again in succeeding months that scientists should keep their chins up, emphasize useful, make their case. Five days after Trump became President, Freilich went to another town hall meeting at the American Meteorological Society. Keep doing what you are doing, he said to the scientists attending. He explained that no one in the new administration had told him to make any changes in the existing program.[2]

What Freilich said to the scientific community, he stated to his staff. They were career government employees, part of "the Deep State" Trump disparaged. As Moore observed: "When Trump was elected, a lot of Freilich's team wanted to jump off the roof. They were fearful, discouraged." Freilich told them "he would go to the front of the line. Take me. I will protect you." He kept up the morale of those under him at NASA, as well as his academic scientific constituency. Everyone knew Earth Science would be a potential target.[3] The sense of threat was palpable.

NASA Leadership

The Trump Administration took a long time to fill its presidential appointments, and this fact especially applied to NASA. NASA's top civil servant, Associate Administrator Robert Lightfoot, served as Acting Administrator during the period when a permanent administrator was absent. His tenure as Acting Administrator extended to a record-setting 15 months. He gave most attention to human space flight. But he was supportive of other programs, including Earth Science, and was quite aware of his need to buffer this program from Trump threats.[4] Zurbuchen remained as head of the Science Division and Freilich Earth Science.

The First Trump Budget

Lightfoot and Freilich both sought to head off major reductions to NASA's Earth science budget in conversations with the Office of Management and Budget. Lightfoot emphasized that any cuts go to satellites early in development.[5] Freilich knew there would be attempted cuts, but his strategy was to have "budget stability." Like Lightfoot, he bargained with civil servants at OMB he had gotten to know over the years as to which satellites to end and which to leave alone. His hope was to get the money back from Congress. Paul Shawcross, Chief of the Science and Space Branch of OMB, exceedingly powerful in budget matters, believed in the importance of Earth Science. He proved a sagacious ally.[6]

On March 16, the White House announced Trump's first budget proposal. The good news was that Trump wanted to use space to project his theme of "making America great again." The budget called for NASA to get $19.1 billion, allowing it to escape the huge cuts inflicted on EPA and other agencies in disfavor. The bad news was that OMB killed five specific Earth satellites it associated with climate change. The satellites for sea-level rise, including those to research polar ice-melt, were spared. Sentinel 6/Jason CS, ICESAT-2, GRACE Follow-On, and SWOT—all survived.

ICESAT-2 and GRACE Follow-On were scheduled for launch in 2018, Sentinel 6/Jason CS for 2020. SWOT, the most expensive and potentially the most endangered, and the heart of NASA's new technology ambitions for sea-level rise, had been approved for final design and fabrication in October 2016. With NASA spending most of the money for the billion-dollar-plus project, it was a likely target. NASA had hoped to launch it in 2020, but this possibility was proving too optimistic. Aside from being well along in development, its linkage with France was important in its protection. Freilich's strategy for building alliances behind technology was not only significant for "offense"—i.e., getting the satellites built—but also for "defense"—keeping them from being excised.

Strengthening NASA's Defense

NASA's defense strategy included not only reinforcing but expanding the long-standing alliance with France. On June 19, 2017, the leaders of NASA (Lightfoot) and CNES (Jean-Yves Le Gall) met and reaffirmed their agencies' decades-long commitment to collaborate, including in oceanography.

In October, the rhetorical strategy (emphasizing uses other than climate change) was on display at a Senate hearing on appropriations. The University Corporation for Atmospheric Research promoted SWOT as an exciting project with "promising uses for flood and drought management at local, regional, and national levels; improved risk assessment by the insurance industry; harnessing ocean energy opportunities; and optimizing the efficiencies and effectiveness of both military and commercial marine operations."[7]

As he had with civil servants at OMB (and Office of Science and Technology Policy), Freilich had nurtured relations with the legislative staff on committees relevant to space, both authorizing and appropriations. Getting to know staff was important as lawmakers changed. Nelson was still a Florida Senator and ranking democrat on NASA's authorization committee. Senator Mikulski, however, retired in January 2017. Republicans now were in charge of both chambers. But many staffers remained. Significantly, the top staffer for Mikulski moved to work with her successor, Senator Richard Shelby (R, Alabama). Freilich altered his message to suit the republicans. The general view of both democrats and republicans on committees was that as long as Earth Science did not impinge on their priorities, which usually had to do with human space flight, they supported it and they did not get into specific satellites.[8]

The attempt to kill named satellites by Trump's Office of Management and Budget displeased many legislators. There was enough about these satellites in terms of uses other than climate change that persuaded conservative lawmakers to retain them. Neither the White House nor Congress went specifically after sea-level satellites. Freilich told Congress he was implementing the agenda of the smartest Earth scientists in the country represented by the National Research Council, not the Obama or any other political agenda. Meanwhile, the NRC worked on its new Decadal Survey for 2017–2027. On a personal level, Freilich had a credible reputation as an apolitical science manager and communicator.

What did not help was a scientific debate that broke into public visibility in 2017. This was called the "Ajustocene" affair. It related to the fact that some scientists, looking at older satellite data, questioned whether the rise was what Nerem and his colleagues said it was—maybe no rise at all! Nerem and his associates recalculated findings and found issues with TOPEX/Poseidon, in particular, the calibration of a sensor. When they made adjustments, they confirmed the overall scientific consensus about what was happening. Nerem said sea-level rise was from around 1.8 mm

(0.07 inches) per year in 1993 to about 3.9 mm (0.15 inches) in 2017. The debate was technical, arcane, and not unusual in science involving a complex and relatively new technology like radar altimetry.[9] The dispute eventually petered out. What mattered most for policy was that any scientific debate about climate change and its impacts added fuel to the political controversy.[10]

The Decadal Survey

At the end of 2017, the NRC Decadal Survey panel delivered its report to NASA.[11] Published shortly thereafter in 2018, it conveyed the scientists' priorities for the next ten years. Freilich had asked the panel not to recommend specific satellites as the 2007 report had done, but instead focus on science issues and questions needing solution.

This survey was quite different from the earlier report. It did indeed focus on science issues. It was also more realistic about costs. The first had an agenda for new satellites that assumed a 30% raise in program budget. That did not happen, although NASA's budget had risen incrementally. It also had estimates for satellite costs that were unrealistic. The result was that many new starts it had wanted were largely still in development in 2018—as SWOT exemplified. NRC now used an independent auditor to assess costs of its recommendations. The emphasis was on finishing what NASA had started before initiating a host of new programs. The earlier decadal was to some extent a "wish list." This later one was an endorsement for the path NASA was on and directive to continue.

Another difference was its greater explicit emphasis on societal needs. Additionally, it discussed management issues, with more attention to the requirement for "continuing research." It noted the tensions between providing scientists with opportunities to extend measurements to new areas and continuing measurements into existing areas. This was a tension that NASA had to balance.

The greater and more explicit attention to climate change and sea-level rise reflected the degree to which this problem had risen on the national and global agenda since 2007. The report mentioned research questions that had to be addressed: How much would sea-level rise, globally and regionally, over the next decade and beyond? What would be the role of ice sheets and ocean heat storage? How would local sea-level change along coastlines around the world in the next decade to century?

It pointed out that new research indicated that the rise from melting ice now exceeded that from thermal expansion. Sea-level rise put the 146 million people worldwide living along coasts at increasing risk. In spite of administration and congressional efforts to clarify agency roles, it complained of continuing "confusion." This fact contrasted with Europe. It pointed out that the EU had formally committed in 2014 to Copernicus, which it called a long-term user-driven program for Earth observation and monitoring. This was "a commitment not just of a nation, but by the EU." Copernicus was a major operational program and the United States had "no comprehensive equivalent." Until the United States made a national policy decision akin to what Europe had done, there was a potential "Valley of Death" between research and application.

Freilich felt he had already been responsive to the "Valley of Death" issue through linking NASA more closely to Europe. Sentinel 6/Jason CS marked the first time that the EU and ESA had partnered with NASA on earth observation. He had connected to operations, while protecting NASA's "distinctive competence" in pushing the technological and scientific frontier. "We did not want to be a victim of our own success," he reflected.[12]

Freilich agreed with the NRC panel on most matters, and told it so. He called the NRC report another Bible to guide the next steps in his program. In 2018, NASA moved ahead with its mission to Earth, in spite of Trump. The president's proposed budget for FY2019 again killed the five satellites targeted the previous year. Congress did not go along once again. The same ritual of cutting and restoring would continue through the four-year Trump term. Freilich knew he could not get a raise for his program. His strategy was geared to defense, with a goal of "budget stability," while planning for the future using the new Decadal Survey.

The Sea-Level Change Team

In March, NASA reconstructed its Sea-Level Change Team (N-SLCT). The team had 50 scientists, one-third new. Leadership shifted from Nerem to Ben Hamlington of JPL. Lindstrom decided to retire at the end of 2018 after 21 years at NASA. He had given his life, he said, to observing the ocean from space. A new program scientist, Nadya Vinogradova Shiffer came aboard to assist and then take over his position. A physical oceanographer, she had worked on Jason missions and been a member of the first sea-level change team and regarded science communication as an important part of the Earth Science program. She made it a priority.

Hamlington specifically urged his team to reach out to coastal planners, to get climate change/sea-level science into their missions when they made relevant findings.[13] Scientists were users of sea-level science, but so also were local and regional officials. Their perspectives reflected practical needs. NASA was now advancing satellite technology so it could better answer research questions as well as project coastal impacts of rising seas. Connecting with coastal users would require developing science communication techniques and making science "usable." Hamlington subsequently enlarged the team, which included both established scientists and their graduate students. He also created an advisory group of practitioners to help put NASA science to work and shifted the emphasis of NASA's Sea-Level Change Team in the direction of projections of local and regional impact.[14]

Bridenstine Takes Command

On April 12, James Bridenstine finally was confirmed as NASA Administrator.[15] After months of waiting, he gained a 51--49 positive vote in the senate. It had been the most contentious confirmation in NASA's history and revealed the deep partisan divide in the Senate. There were many reasons lawmakers opposed Bridenstine. One was his stance while an Oklahoma Congressman against climate change science. He had come across as a denier.

But now he stated his views had "evolved." Burned by his own problems in getting confirmed, he vowed he would do everything in his power to keep NASA "nonpartisan" and above the Washington fray. That included climate change politics.

On May 12, he came to NASA's auditorium and addressed NASA's employees. Thousands of others could listen via links to NASA field centers. He declared that the climate was changing and that humans were a cause. He said he supported NASA's research in this field. NASA, he declared, was "the one agency on the face of the planet that had the most credibility to do the science necessary so that we can understand it better than ever before."

He said NASA would follow the NRC guidelines, and he believed climate change research was essential and NASA would reduce "climate uncertainties." In contrast to others in the Trump Administration, he said climate science was NASA's mission and indicated it would not be moved elsewhere.[16] His remarks relieved many in NASA and its science constituency who had been worried about a shift of NASA Earth Science to NOAA—*without* more money to NOAA. They felt that would be the road to termination.

Launching

The two GRACE satellites went up May 22. The first GRACE satellites had proved quite effective, and GRACE Follow-On satellites were expected to be better. A month later on June 20, NASA and European partners celebrated the tenth anniversary of Jason-2. EUMETSAT declared that Jason-2 had paved the way for the movement of ocean satellite technology from research to operations, a process reaching fruition with Jason-3 and its successors. Josh Willis, a leading ocean scientist at JPL, noted that Jason-2 would "keep us on the pulse of climate change."[17]

The big pay-off was ahead, various observers believed, with SWOT, whose sensors would better connect the oceans with coasts and inland seas. In June, France announced it had made substantial progress on its part of the project. Indicative of the significance France accorded SWOT, that mission was receiving resources under a strategic fund France had established called its "Investment in the Future Plan."

Freilich Decides to Leave

Freilich was beginning to think that it might be time to retire. He was soon going to be 65, had been working extremely hard to reinvigorate the Earth Science program since coming aboard in 2006, and the Trump years were stressful and enervating even though his program was still holding its own. His wife wanted to return to Oregon. He had his staff converting the new Decadal Survey into program plans. A successor could start with the Decadal Survey and this roadmap. It seemed the right time.

On August 29 he announced he would leave the next year, 2019. When he signed the order that made a launch of ICESAT-2 possible, he "wept openly," knowing this was his "last launch as Earth Science Director."[18] Two weeks later ICESAT-2 roared into space. This launch was particularly important for NASA (and for Freilich). This was not only because it had to do with glacial melting, but because its development had been technically troubled. The cost was up to $1 billion. Freilich had kept studies of ice melting at the poles going through ICEbridge, but airplane observation could not do what ICESAT-2 could do. ICEbridge ended as ICESAT-2 began.

As ICESAT-2 launched, an emotional Freilich looked on. What he knew about the loss of ice, especially in Greenland, was alarming. The process was accelerating, and the waters around it were rising faster. That made ICESAT-2 particularly important. Initiating it had been one of his first decisions as Earth Science's leader. He had come full circle with its deployment.

Jason-3

In January 2019 Jason-3 completed its three-year primary mission, with all equipment going well. NASA believed it would last several years beyond its planned five-year schedule. This was an operational satellite, and EUMETSAT spokesman, Renko Scharroo, called Jason-3 "crucial for climate monitoring." Not only had the Jason series given years of data establishing the reality of rising seas, he said, but now the first signs of acceleration of rise had been observed.[19] This finding would not have been credible without data continuity, a goal that had united all involved with the program.

CNES stated that while Jason-3 was the latest of a series of largely duplicate satellites, it did incorporate new technology that added to its policy relevance. It pioneered a better scan of coastal areas, lakes, and rivers. Previously, satellites had not covered these critical places, with "significant socioeconomic and climatic issues," well.[20]

The international partnership behind Jason-3 and other satellites was debated in Congress in April, particularly in connection with the upcoming satellites—those called Sentinel 6A and 6B/Jason-CS. Some lawmakers thought NASA was providing too much support for European satellites. Why? And what about the growing commercial satellite sector? Was NASA competing? Bridenstine and others fended off the queries.

The Rhythm of Innovation

There was also criticism of NASA overruns generally from Congress, and these hit the Earth Sciences Program primarily through ICESAT-2. Critics questioned NASA on ICESAT-2 in May and June hearings. Bridenstine again defended his agency and its programs.

There was a rhythm to what was happening—as certain satellites aged and left the scene, others came on and replaced them. The key issue for the program in 2006, when Freilich arrived, had been the absence of satellite successors in the future. That was still a challenge, developing the new to keep data continuity of measurements, but Freilich had prevented the "collapse" that worried the Earth science community earlier. Sea-level rise was a particularly robust mission, with research and operational fronts, and a team of scientists to help provide advice to NASA and outreach to users.

Part of this new momentum was the strong partnership with Europe and NASA's leverage of Europe's expanded program for its own interests. As fears of Trump cuts dimmed, change and renewal had become almost expected. Jason-3 was going strong as Jason-2 came to an end in October. Two more Jasons were being built in a merger with Europe's Sentinels, guaranteeing stability throughout the 2020s. And, perhaps most important for NASA, SWOT development, although much delayed, moved forward, the first of a new generation of ocean-observing satellites. It would significantly track change in all the Earth's waters. When ready, NASA and its long-term partner, CNES, would help make SWOT part of the climate infrastructure observing the planet.

Behind the sequence of satellites, exerting a pull from policymakers was the knowledge the satellites enabled. Sea-level rise was speeding up, and so was the melting of ice at the poles. If there was some residual debate about Antarctica, how much melt and where, there was little to none about Greenland. GRACE as well as ICESAT showed the loss of ice was real. Warming at the poles was more intense than elsewhere on the planet. Josh Willis of JPL called sea-level rise "the most complete measure of how humans are changing the climate."[21]

An Administrator's Legacy

When he announced his retirement in late 2018, Freilich was asked what he would do next. He replied that he and his wife wanted to "travel and explore the planet we committed to understand and protect."[22]

Soon after he retired in February 2019, Freilich was diagnosed with pancreatic cancer, a fatal disease. When Josef Aschbacher, a close friend who headed ESA's Earth Observation Program—and who would eventually be director of ESA—heard, he wanted to make sure Freilich got the recognition he deserved, before he passed. He spoke with EUMETSAT and other European colleagues, and they agreed Sentinel 6/Jason-CS would be renamed. It would be called Sentinel-6 Michael Freilich. It was rare to have a satellite named after an individual and particularly a government administrator. But there was agreement on both sides of the Atlantic that he deserved the honor.

On January 28, 2020, the Europeans and NASA held a ceremony at NASA to officially recognize the name change and honor Freilich, who was able to attend. Aschbacher declared that "this mission demonstrates what NASA and ESA can achieve as equal partners in such a large space project." He pointed out that adding Freilich's name to Sentinel 6 "is an expression of how thankful we are to Mike. Without him, this mission as it is today would not have been possible."[23] Freilich's remarks echoed those of Aschbacher and others who emphasized the criticality of working together to study the Earth and its ills. Many at the ceremony commented on the difficulties of making bureaucracy work and collaboration across nations. What made Freilich stand out, they agreed, was his ability to make government effective. On August 3 Freilich died. Bridenstine stated that the Earth had "lost a true champion."[24]

Sentinel-6 Michael Freilich sailed into space on November 21, 2020, from Vandenberg Air Force Base in California. With a twin already authorized for a 2025 launch, these satellites marked the beginning of the institutionalization of sea-level rise observations. It would take more time to complete a process of large-scale, long-term change in government. But the process was well underway with a transatlantic base of support and progress. The new satellites would extend the sea-level record begun by TOPEX/Poseidon in 1992 another decade, and almost surely beyond. Freilich's close family and friends attended the launch and Aschbacher again paid tribute to Freilich, noting how much he championed international collaboration to confront the challenge of sea-level rise.

Notes

1. Debra Werner, "Earth Scientists are Freaking Out: NASA Urges Calm," *Space News*, (Dec. 28, 2016).
2. Ibid.
3. Berrien Moore, interview by author, Apr. 14, 2022.
4. W. Henry Lambright, "Maintaining Momentum: Robert Lightfoot as NASA's Acting Administrator, 2017–2018," *Space Policy* (Apr. 2019).
5. Ibid.

6. Charles Bolden, interview by author, Dec. 27, 2018.
7. "Commerce, Justice, Science, and Related Agencies Appropriations for Fiscal Year 2018." *GovInfo | U.S. Government Publishing Office,* (2017). Retrieved from https://www.congress.gov/event/115th-congress/senate-event/LC49303/text
8. Freilich interview by author, Feb. 18, 2020.
9. Josh Willis, interview by author, March 16, 2023.
10. Jeff Tollefson, "Satellite Snafu Masked True Sea-Level Rise for Decades," *Nature 547*, 265–266.
11. NRC Decadal Survey, *Thriving on Our Changing Planet: A Decadal Survey for Earth Observation from Space*, (2018).
12. Michael Freilich, interview by author, Feb. 18, 2020.
13. Pat Brennan, "Teamwork: NASA Sea-Level Scientists Join Forces," *NASA's Sea Level Portal.* Retrieved from https://sealevel.nasa.gov/news/116/teamwork-nasa-sea-level-scientists-join-forces/
14. Ben Hamlington, interview by author, Apr. 18, 2023.
15. Marcia Smith, "Bridenstine Confirmed as NASA Administrator on Party-Line Vote," *SpacePolicyOnline*, (Apr. 19, 2018).
16. Sarah Lewin, "New NASA Chief Bridenstine Says Humans Contribute to Climate Change in a Major Way," *Space.com*, (May 19, 2018).
17. Alan Buis., "Prolific Sea Observing Satellite Turns 10," *JPL News*, (June 20, 2018). Retrieved from https://www.jpl.nasa.gov/news/prolific-sea-observing-satellite-turns-10
18. "Statement of NASA Administrator James Bridenstine," *NASA Blogs*, (Aug. 5, 2020). Retrieved from https://blogs.nasa.gov/bridenstine/2020/
19. "Jason-3 Altimetry Mission," (Jul. 24, 2013). Retrieved from https://www.eoportal.org/satellite-missions/jason-3#mission-status
20. Ibid.
21. Arielle Samuelson, "New Earth Mission Will Track Rising Oceans Into 2030," *NASA*, (Nov. 20, 2019). Retrieved from https://www.nasa.gov/feature/jpl/new-earth-mission-will-track-rising-oceans-into-2030
22. Paul Voosen, "NASA's Long-Serving Climate Chief to Retire Next Year," *Science*, (Aug. 20, 2018). Retrieved from https://www.science.org/content/article/nasa-s-long-serving-climate-chief-retire-next-year
23. Karen Graham, "Sentinel-6 satellite renamed in Honor of Renowned U.S. Scientist," *Digital Journal*, (Jan. 28, 2020). Retrieved from https://tinyurl.com/22vmtw4s
24. "The NASA Family Mourns the Loss of Dr. Mike Freilich – NASA Administrator Jim Bridenstine." *NASA Blogs*, (Aug. 5, 2020). https://blogs.nasa.gov/bridenstine/2020/

CHAPTER 12

Advancing

Abstract This chapter covers the first two years of the Biden Administration (2021–2022). Biden unshackled NASA and increased its Earth Sciences budget beyond the $2 billion mark. He approved a new five-satellite program for the 2020s—in line with the 2017–2027 NRC Decadal Survey. NASA plans included further study of glacial melt and sea-level rise. NASA spoke out more strongly about the threat and enhanced its sea-level change team's outreach activity with an aim for climate projections. SWOT, a Freilich legacy and $1.2 billion next-generation satellite with CNES that looked not only at the ocean but inland seas, launched in late 2022. NASA and the European Space Agency (ESA) agreed to work closely on further satellite development. ESA was willing to shoulder much of the operations NASA wished to avoid in favor of pushing advanced technology and scientific breakthroughs. Working out this balance between R&D and long-term monitoring, as well as relationships among agencies to further satellite development, would be up to NASA's new Earth Sciences Director, Karen St. Germain, an electrical engineer who had headed NOAA's weather satellite activity.

Keywords European Space Agency (ESA) • Karen St. Germain • Joseph (Joe) Biden

W. H. Lambright, *NASA and the Politics of Climate Research*, Palgrave Studies in the History of Science and Technology, https://doi.org/10.1007/978-3-031-40363-7_12

Sentinel-6 Michael Freilich was part of the TOPEX/Poseidon/Jason series and carried improvements based on specific requirements. Jason-3 had advances that entailed better views of coastal impacts of rising seas, and so would this one. It would be able to gauge ocean height within 984 feet (300 meters) of the coastline. This was much closer than earlier Jasons. As Paul Voosen wrote in *Science*, "The coasts are where sea-level rise hits home."[1]

And sea-level rise was hitting home more seriously. IPCC, using NASA and other data, had stated in a 2014 report that seas were rising 3.2 millimeters (0.126 inches) every year. But more recent data, according to the chair of NASA's sea-level change team, Ben Hamlington, now showed larger increments—4.8 millimeters (0.189 inches) a year.[2]

This increase was serious. The rise was also inexorable, fed by ever larger ice-melt and ocean warming. It would, in a matter of decades, reshape the coasts. Floods that came infrequently would occur annually in vulnerable places. There were more challenges ahead, and NASA's Sea-Level Change Team developed a computer-based portal that provided updated information to a multitude of users.

Freilich's departure and death did not slow the satellite program. His Deputy, Sandra Cauffman, maintained the momentum of planning he had initiated in the wake of the Decadal Survey. She held sway for 17 months while NASA looked for a successor. On June 8, 2020, Karen St. Germain became NASA's new Director of Earth Science. She had a different background from Freilich, with a PhD in electrical engineering. She previously had headed NOAA's satellite activity, including the Joint Polar Satellite System.

Unshackling NASA

St. Germain faced a political environment for climate change and sea-level rise that shifted dramatically when Joe Biden became President in January 2021. He immediately made climate change a government-wide priority.

Having been Obama's Vice President, Biden was thoroughly familiar with the climate change issue and had pledged to make climate change a signature part of his agenda if elected. On his first day in office, January 20, 2021, he returned the United States to the Paris Climate Accord that Trump had left and began signing executive orders rescinding the rollbacks that Trump had instituted of environmental regulations. He soon made appointments of climate staff and advisers and created an inter-agency task force to elevate climate change across the government.

NASA was not initially on this inter-agency body, but the agency insisted that science—and it—be at the table.[3] Even before NASA had a new permanent Administrator, the acting Administrator, Steve Jurczyk, appointed Gavin Schmidt, who held the job Hansen had had of Director of the Goddard Institute for Space Studies, to be a high level spokesman for the agency on climate science and technology. Coming across as less strident than Hansen, he was direct, clear, and capable of communicating that climate change was a crisis requiring action.

On May 3, Bill Nelson, former Florida Senator, and long-time NASA enthusiast, was confirmed as Administrator. While most concerned with human space flight and returning to the Moon, he was extremely supportive of making Earth Science and climate change a high NASA priority. As a Florida Senator, Nelson had had hearings dealing specifically with the sea-level threat. Now, as NASA Administrator, he called NASA "the point of the spear in climate change.... the leading climate agency."[4] The new Earth Science Director, St. Germain, faced a political environment that was not only supportive but anxious for NASA to be more proactive and visible on the climate front. NASA's Earth Science budget soon gained substantially in accord with Biden policy.

The Earth System Observatory

As one of his last actions before retiring, Freilich had started plans to follow-up on the 2017–2027 NASA Decadal Survey. The plans evolved under Cauffman, took greater form, were finalized, and got a name under St. Germain. An initiative came to light officially in Biden's first budget, which was proposed in early June 2021. With Earth Science cracking $2 billion, it included a decision to spend $2.5 billion over the ensuing ten years for five new spacecraft to monitor the changing climate. The five were proposed as an integrated Earth System Observatory (ESO). The ESO would get $137.8 million in its first year for preliminary work and then scale up as development proceeded. As the new cluster advanced, so would the budget of the overall Earth Science program.

"It's pretty exciting," Nelson stated. The new program "is going to measure all the intersections of the atmosphere, land, the ice, and the oceans. And it's going to do this over the course of the next decade."[5] St Germain declared: "Earth system science is poised to make an enormous difference in the ability to mitigate, adapt, and plan for changes we're seeing. The pace we're going to have to do that is much higher in the decade in front of us than the decade behind us."[6] She indicated that she didn't view the Earth System Observatory "as a menu of systems you can select from." She declared: "I view this as an integrated observatory, where we are looking at all the major 'muscle movements' of the Earth system."[7]

The Era Ahead

One of the five new satellites planned for later in the decade was a possible successor to GRACE. GRACE Follow-On had been a bridge, based on what NASA could afford. It was deemed essential to monitor the melting of glaciers and transfer of mass to ocean waters. Outside the new five, but closely related, was SWOT. It was increasingly imperative to understand, predict, and mitigate the danger of sea-level impacts on localities and regions. SWOT and other new tools would improve capability for disaster warnings. With such considerations in mind, Canada and the United Kingdom joined NASA and CNES as contributing partners in the SWOT project.

After he retired, Freilich had told Moore he wished he had given applications "more attention" when he had managed Earth Science. His emphasis as Earth Science leader was science-building, creating new capability. What exactly he meant was not clear, but he had edged in the (operations) direction with his arrangement with Europe on the satellite named after him. It seemed logical that like disaster mitigation would get more priority in the future under Biden, but it was not a given that NASA would take the lead in this area. While NASA dominated in climate satellites, NOAA also faced an environment in which it could be more active in ocean science and getting science applied. The more NASA moved into communicating usable science, the more it would potentially impinge on NOAA's turf. Having an Earth Science leader who was from NOAA presumably would help to mollify jurisdictional rivalry.

NOAA had been traditionally active in coastal matters. The coasts were where most of the damage from sea-level rise would take place. Tide gauges were good for measuring sea-level at coasts, but satellites were increasingly critical for connecting the ocean with coastal problems. Moreover, scientific studies showed agreement between the two technologies. Significantly, the newest satellites were getting closer to observing the coasts in important detail. The most recent satellite (SWOT) would get "very close" to the coasts.[8] Local sea level was seen as a growing issue in science, technology, and public policy from a standpoint of coastal resilience. NASA would probably have to evolve a greater coastal presence.

An Expanded Partnership

On July 13, 2021, NASA and ESA signed an agreement on climate science cooperation. The aim was to create an overarching strategic framework that would expand on the current NASA–ESA arrangement for Sentinel-6 Michael Freilich. Ausbacher, now head of ESA, and Nelson hailed the agreement and its import for the future. Nelson declared: "Climate change is an all-hands-on-deck, global challenge that requires action now. This agreement will set the standard for future international collaboration, providing the information that is so essential to tackling the challenges posed by climate change and helping to answer and address the most pressing questions in Earth Science, for the US, Europe, and the world."[9]

ESA, with 22 member-states, planned to spend $16.9 billion euros ($17.5 billion dollars) in 2023–2025 on a range of projects. Copernicus was a priority. NASA intended to continue emphasizing cutting-edge research and development, and ESA evinced willingness to conduct "long-term" research (i.e., monitoring). That linkage (NASA–ESA) gave promise for the future.

So also did the launch of SWOT December 16, 2022. Deploying the substantially advanced capability of wide-swath technology, the $1.2 billion satellite marked a transition for the TOPEX/Poseidon/Jason series to a new generation of satellites. The improvement in resolution was 10-fold. Scientists could thereby better observe, understand, and predict how perturbations in the open ocean could cause sea-rise havoc on the mainland.[10]

SWOT would "zoom in on our coasts," declared Nadya Vinogradova Shiffer, SWOT scientist and program manager at NASA. This meant it could "bring into focus" not only sea-level rises but its impact on flooding and water quality so critical to the public.[11] NASA planned for follow-ons to existing sea-level satellites that would build on what had been launched.

Also, NASA explored a stronger effort to assist vulnerable "communities prepare for the effects of the Earth's Rising ocean." While NASA had worked with users in the past, there was a greater sense of urgency to translate science into policy under Biden.[12] NASA was pushing its interdisciplinary sea-level change team to integrate knowledge from sea-level and other satellites into an integrated usable science for coastal regions affected by climate change. NASA had built an infrastructure of satellites, and now it had to be put to work for human resilience. It had taken a long and tortured journey to get to this point, but that multi-decade journey made next steps possible.

Notes

1. Paul Voosen, "Seas are Rising Faster Than Ever," *Science* (Nov. 26, 2020), 901.
2. Ibid.
3. Alexandra Witze, "NASA Reboots Its Role in Fighting Climate Change, *Nature*, (May 7, 2021).
4. Jeff Foust, "Harris says National Space Council will Develop 'comprehensive framework' for space priorities," *SpaceNews* (Nov. 6, 2021).
5. Jeff Foust, "An Aggressive Budget for More Than Just Earth Sciences," *Space Review*, (June 1, 2021). Retrieved from https://www.thespacereview.com/article/4186/1
6. Paul Voosen, "NASA Set to Announce Earth System Observatory," *Science*, (May 7, 2021), 554–555.
7. Jeff Foust, "NASA brands future Earth science missions as Earth System Observatory," *Space News* (May 25, 2021).
8. Josh Willis, correspondence with author, Mar. 22, 2023.
9. Jeff Foust, "NASA and ESA Sign Agreement on Climate Science Cooperation," *SpaceNews* (July 14, 2021). Retrieved from https://spacenews.com/nasa-and-esa-sign-agreement-on-climate-science-cooperation/
10. Paul Voosen, "NASA Radar Altimetry Mission to Study Hidden Ocean Swirls," *Science* (Dec. 9, 2022), 1032; See also Josh Diner, "SpaceX Launched NASA Satellite to Study World's Water." *Space.com* (Dec. 16, 2022).
11. Pat Brennan, "SWOT Satellite: Bringing Earth 's Coastlines into focus," NASA, Dec. 13, 2022. Retrieved from sealevel.nasa.gov.
12. Nadya Vinogradova and Benjamin Hamlington, "Sea Level Science and Applications Support Coastal Resilience," *EOS*, (June 29, 2022).

CHAPTER 13

Conclusion

Abstract This concluding chapter calls the NASA Earth Science/sea-level rise program a success. There was success owing to sustained bureaucracy-driven policy in spite of an often chaotic and sometimes hostile national context for climate change. This was a successful program in going from vision to transatlantic commitment because of: (1) unifying goals—both long-range and short-term; (2) a supportive constituency; and (3) dedicated administrative leadership. Leaders used various strategies, including international partnership, to gain their ends domestically. What stands out is human and organizational persistence over decades coupled with a dose of shrewdness.

NASA's sea-level rise program is likely to continue and advance as long as the danger of climate change endures—a very long time. There are plans for successors to current satellites. What the past shows is that NASA will need to adapt its research program—and risk communication—not only to an evolving geophysical problem but also to an ever-changing political and international environment.

Keywords Bureaucracy-driven policy • Unifying goals • A supportive constituency • Dedicated administrative leadership

W. H. Lambright, *NASA and the Politics of Climate Research*, Palgrave Studies in the History of Science and Technology, https://doi.org/10.1007/978-3-031-40363-7_13

The seas are rising. The rise is quickening. The danger is increasing. How do we know? One major reason we know is the advent and use of space satellites, along with other technologies, such as tide gauges, and related research. Satellites give scientists the ability to view the vast expanse of oceans in remarkable detail. They allow specialists to detect not only how much the seas are rising but the causes of that rise that are showing impacts on the coasts. Aside from thermal expansion, a growing reason for accelerated rise is the melting of glaciers and ice sheets at the poles. Behind this knowledge are decades of uninterrupted observations from a government program devoted to this problem. This book has been a policy history of that U.S. government program, which is under the authority of NASA. This means that one of the most important consequences of the space program has been the critical knowledge it gives us of planet Earth.

Policy Innovation

The approach of this book has been to study this program as a long-term process of policy change that has gone from initiation to institutionalization. The process of policy/program innovation is just as important as the technological innovation of satellites. It enables scientists and engineers to do their vital work. The program was once new and required risk-taking entrepreneurship to get started. It has required continued nurturing and strategizing since. It has gone from awareness of possibility in the 1960s to routinization in the 2020s. What exists today is an infrastructure of operational satellites with the expectation of continuous improvement through research and development. The program behind this system is likely to continue as long as climate change remains a major driver—likely a very long time.

This is a book about government and science in America. As a study of federal policy, it has its locus at the administrative level of government. Agencies are theoretically and practically implementers of policies made by national political institutions. But what if national political institutions do not make policy? Or make it poorly? Agencies can do nothing or act. To the extent they get national political institutions—the President and Congress—to acquiesce, they make policy, or at least influence it significantly. In the absence of coherent and overarching national policy for climate change, including that for climate change science, NASA has moved ahead and forged a program, a specific mission. And that NASA program of satellites and related research has helped yield a base of knowledge about one of the most compelling impacts of climate change, namely sea-level rise.

This program is a classic example of incremental bureaucratic policy-making, evolving over years in response to various challenges. It has been cobbled together by a sequence of administrative leaders. It began in the 1970s when sea-level rise was strictly a matter of speculation, with Seasat. That was a demonstration of technology capability, "spin-offs," in NASA parlance. How could space technology be used for observing the seas? Narrowly oriented, it revealed important benefits of viewing the ocean from space in 1978. Failure of Seasat after a brief life triggered NASA in 1979 to forge an official "ocean-from-space" program that embraced both satellite development and research. The initial purpose was understanding of ocean dynamics, particularly massive currents ship-based research could not view.

To get at that understanding, NASA sought to develop a more sophisticated satellite called TOPEX. The need for funds led to collaboration with the French and merger of programs to produce TOPEX/Poseidon. In 1992 TOPEX/Poseidon launched and performed with a precision that amazed most of those involved. It enabled researchers to discern the rise of oceans due to climate change.

TOPEX/Poseidon was a breakthrough. It marked the beginning of a sea-level rise focus within NASA. It became a major component of a larger multi-faceted Earth observation program that came later. TOPEX/Poseidon was followed by a sequence of satellites called Jason-1, Jason-2, Jason-3, and Sentinel-6 Michael Freilich in 2020. Another sea-level satellite akin to Sentinel-6/Michael Freilich is scheduled for later in the decade, and a second-generation satellite, SWOT, launched in late 2022.

There have thus been continuing measurements of sea-level rise since 1992. That continuity over decades has allowed scientists to know how much rise is happening and how fast. Other satellites have investigated causes, particularly the melting of gigantic ice formations in Greenland and Antarctica. Another cause, observable by NASA from space through AQUA and also under the ocean surface by a NOAA system called ARGO, tracts thermal expansion. Together, and with local tide gauges, there is a technological infrastructure for sea-level rise enabling critical scientific research and public policy.

This study concentrates on the NASA role, which has been catalytic in creating a system of satellites. The technological system required a coalition of agencies on both sides of the Atlantic. While contested, given climate change politics, NASA's program has taken sea-level rise from a concept foreseen by few to an institutionalized reality used by many in the 2020s. That is a mark of success and shows government can work in spite of innumerable hurdles and opposition.

What were critical factors shaping this NASA program? While there were many, three overlapping factors stand out contributing to success. There was a unifying goal, a supportive constituency, and administrative leadership.

A Unifying Goal

Since TOPEX/Poseidon, sea-level rise satellites have been associated with broader observational efforts at NASA managed by its Earth Science Division. These collective activities had aspirational goals such as building an "Earth System Science," or "Science of the Earth," or "Integrated Program," "or a "Predictive Capability." "Understanding Climate Change" has also been a goal, but often understated for political reasons. Such high-level, even noble goals are desirable as beacons keeping large and diverse groups united. But they are not enough.

What has also particularly unified those connected with sea-level rise has been an interim or "working goal" that resonated with NASA, other agencies, and scientists. This was "data continuity." Once TOPEX/Poseidon, using altimetry instruments, made its breakthrough in 1992, program advocates seized on maintaining a long, continuous, ultra-precise record. Only with a credible record could scientists be sure of what was happening to the sea's changes in height and the pace of rise. One-shot affairs were not sufficient for NASA or key users interested in the impacts of climate change, such as IPCC. It was exceedingly difficult to maintain a steady sequence of satellites in a governmental system geared to annual appropriations and stifled by polarized politics.

As NASA launched one satellite, advocates lobbied for a successor. There was data continuity from TOPEX/Poseidon to Jason-1 to Jason-2 to Jason-3 to Sentinel-6 Michael Freilich and others. That made this program a testament to persistence and ingenuity of those involved in its promotion and sustainment. Given the fact that many government programs fall by the wayside in implementation, this continuity over many years is as important as it is unusual.

"Data continuity" is not an elegant, inspirational goal, such as "Earth System Science," or "Science of the Earth," or some others that were used. It is means as well as end. But it made the more inspirational objectives possible. It worked as a unifying, underlying driver for scientists, engineers, and administrators across agencies and nations for decades. For NASA, in particular, it helped connect separate satellites launched when possible into a more coherent and inclusive program that emerged slowly. However, that program faced innumerable roadblocks. Hence, more was needed to succeed than unifying rhetoric or the luck of satellites with longevity.

A Supportive Constituency

Every government program, to be successful, needs a supportive constituency inside and outside the agency that is its home. For a science and technology satellite program such as sea-level rise, there has been a technical constituency and a broader administrative/political coalition on which it relied for support. The technical constituency consisted at first of remote sensing engineers and oceanographers. Over time, glaciologists came aboard the sea-level coalition. Other fields concerned with understanding the rise and fall of coastal land joined. The problem of climate change/sea-level rise needed an interdisciplinary approach and attracted able people willing to work together, especially on the Decadal Surveys and Sea-Level Change teams. The fact that many in this community depended on NASA funding for research support wedded scientists to NASA in a mutually dependent alliance.

Having a technical constituency of scientists and engineers is not enough. The career administrators at NASA who ran Earth observation and sea-level rise research needed other managers in NASA and the agency in general to provide support. In a large, hierarchical organization like NASA, that meant chiefly the NASA Administrator—the political executive in charge. Then, more broadly, came Congress and the White House, with the president's surrogates in OMB and OSTP particularly significant. A government science program needs legitimacy in legislation and resources in budget. It also needs outside scientists to give technical legitimacy and inside scientist-managers to turn possibilities into programs. Administrators are at the nexus of science and politics. They are also accountable and need allies to get and use resources for implementation. Directly and indirectly, quickly or gradually, effectively or imperfectly, individuals and organizations have to be brought together in a coalition behind a program to make it viable and defend it against threats.

The support of other agencies became essential for this program. The logical domestic partner for sea-level rise was NOAA. Early on, NASA viewed NOAA as the right partner in ocean innovation, based on the weather satellite model. But NOAA's ocean organization was weak in contrast to that for weather. The pull from weather bureaucrats was strong, but NASA found it had to pull NOAA along and play a very different role in oceans than weather. When it came to ocean and climate research, NOAA was budget-constrained and its location in the business-oriented Commerce Department did not help. NASA enlisted NOAA in its constituency, but did not find it a reliable partner.

NASA saw other agencies abroad as more dependable. The French space agency, CNES, proved to be a consistent, long-term constituent, and collaborating partner. Like NASA, it was research-oriented, and it helped NASA administratively, technically, and politically. As the sea-level program expanded, NASA's European constituency grew, with the EU, ESA. EUMETSAT, and others eventually joining. Organizational constituents matter in practical ways, through resources in money, expertise, and political backing. Constituents can be relatively passive—as has been true of many users of satellite data. Or they can be proactive, as particular activist scientists in the United States and CNES abroad have been throughout the program's history. Such constituents as Wunsch in the early years and Moore and others later proved critical to NASA and the shaping of its sea-level program.

At the elected political level, certain legislators have been supportive constituents. Senators Mikulski and Nelson stood out as the most involved and influential politicians in NASA's congressional constituency. Both were genuinely interested in the space program. Mikulski's interest was reinforced by her strong loyalty to NASA's Goddard research center, in her state of Maryland. Having Goddard as one of NASA's two prime centers for sea-level research helped NASA's budget. Nelson's personal interest was elevated by his representation of Florida, arguably the state most vulnerable to sea-level rise.

While those senators were most visible—and most influential as NASA's elected constituents—they were not alone. There were other legislators who were supportive over the years, particularly those from U.S. coastal states. Some Republicans supported sea-level rise research, but avoided the term "climate change." What mattered to NASA was the resource support, not the rhetoric. It was also easier for Congressional politicians to support a smaller program launched incrementally than one of "big science," as shown by the decimation of EOS. Key satellites associated with sea-level rise could fly under the radar of climate politics.

Administrative Leadership

Administrative leadership is the third and most important factor in accounting for success and failure in government programs, especially those that take years to go from an initiative to institutionalization, and where "national policy" is inchoate. It enables and embraces the previous components: identifying unifying goals and building supportive constituencies. It turns a constituency of stakeholders into a political coalition to promote and defend a large-scale technical program.

The leadership for climate science and sea-level rise has come from senior career bureaucrats at NASA. The space agency has expertise and reputation. Add to these ingredients the capacity of its leaders to forge advocacy coalitions of scientists, other agencies and politicians, as well as counteract opponents, and the result is bureaucratic power in public policy. Such bureaucratic influence is obviously limited by democratic accountability. Other institutions check the administrative state, but some agencies, and some career executives, are granted more autonomy than others. Ability does matter, reputation matters, as do results. The much-maligned "Deep State" keeps long-term programs going as politicians and others come and go.

The challenge for any administrative leader is to gain enough autonomy to act and to build a coalition of collaborators and others to reach goals and fend off threats to a program. Luck also matters, in the sense of favorable political environment, but successful leaders take advantage of fortune to make the most of opportunities. They also deal with bad luck—such as negative political change—with determination to do the best possible to limit damage. Windows of opportunity can close as fast as they open.

To be sure, individuals outside government mattered in this case, such as Wunsch in the early days and Moore later. However, federal administrators had to make use of them. They had the governmental power—some more than others—to utilize such allies. Wilson, Kennel, Townsend, Asrar, and Freilich were the prime NASA leaders over the long history of this program. St. Germain, Administrator at the time of writing, begins on the base they built. They have had the right positions for possible influence. As a group they were able to take a new program from birth to institutionalization, making use of a range of strategies. Not all leaders have been equally effective, but as a collective, they have moved the program forward. What have these bureaucratic leaders done? How?

The Initiator: Stanley Wilson, 1979–1992

Wilson did not embark on a targeted sea-level rise program, but he enabled it through TOPEX/Poseidon and the coalition behind it. He started with the NASA Administrator, (Frosch), an oceanographer, on his side. That gave him a mandate and resources to initiate an "ocean-from-space" program, as he called it. He had a degree of autonomy with minimal organizational constraints. He started coalition-building with remote-sensing engineers and oceanographers. Townsend, his deputy, and Wunsch, an MIT scientist, embodied this core technical alliance. He added Goddard and JPL for institutional backing and used an oceanographic association to enlist the broader ocean science community. He won the support of a second NASA Administrator (Beggs) by extending his constituency to an international partner, CNES, to share costs.

His major challenge at NASA in the 1980s was a rival, Tilford, whose big science program, EOS, had more support from NASA. His program, emphasizing TOPEX/Poseidon, was subordinated, although instrumental to EOS and NASA as an EOS precursor. The Challenger shuttle's accident in 1986 was the opening he needed to gain priority and a launch date. However, the most likely future beyond this one launch for TOPEX/Poseidon altimetry was absorption into a concentrated EOS and evisceration of his program's identity. Wilson himself lost autonomy and his authority prior to launch.

However, the vast scale and cost of EOS meant that it could not survive intact unless it became a national priority, and it failed to do that in competition with the Space Station. When Dan Goldin became Administrator, he removed Tilford and ordered EOS downsized, a process already begun. Wilson had departed by the time of TOPEX/Poseidon's successful launch in 1992, but his prime legacy was the sea-level rise program that emerged from the breakthrough that TOPEX/Poseidon represented. Wilson had bureaucratic entrepreneurship in his personality, and it is noteworthy that after NASA, at NOAA, he promoted Jason-3 and established ARGO as a complement to NASA satellites for sea-level rise.

The Remaker: Charles Kennel, 1994–1996

Kennel's prime task was to de-scope and downsize EOS while keeping as much science as possible. He did that, shifting the overall MTPE design from a concentrated to distributed satellite system weighted in favor of three intermediate-scale multi-instrument descendants of the original EOS. That new model made resources available for smaller sea-level satellites to gain independent opportunity within NASA. TOPEX/Poseidon's extraordinary success won the oceanographers' support and stimulated NASA and CNES to proclaim "data continuity" as a clear working goal for Jason-1.

It also opened up funding for another specialized satellite, ICESAT, to measure how the loss of ice in glaciers and ice sheets contributed to sea-level rise. Kennel had Goldin's support for the changes he made. His deputy, Townsend, proved a behind-the-scenes advocate for altimetry, with an enthusiasm and dedication that went back to his pioneering engineering role with Seasat in 1978. Kennel, an astrophysicist from UCLA, relied greatly on Townsend for institutional memory and key decisions.

Although at NASA only two years, Kennel was important in keeping NASA's Earth Science program viable politically as well as technically. With climate change politics growing in congressional controversy during his tenure, he molded NASA's rhetorical strategy for science communication: say what the science dictates, publish, but do not make an alarmist deal of it, especially to the media.

The Consolidator: William Townsend, 1996–1998

Townsend moved from deputy to Acting Associate Administrator of the Mission to Planet Earth division when Kennel left. He continued and consolidated the changes made when he had worked with Kennel. He added another mission to the Jason-1, ICESAT duo, called GRACE. GRACE measured mass, and thus could detect how ice-loss translated into sea-level gain.

Townsend, a career NASA official, operated quietly but highly effectively as a self-proclaimed bureaucratic "survivor" under the volatile Goldin. Most of the technical and organizational changes from the EOS deconstruction, especially those connected with sea-level rise, made it through a turbulent period organizationally inside and politically outside NASA. They were in development as Townsend gave way to his successor.

The Embattled Maintainer: Ghassem Asrar, 1998–2006

Under Ghassem Asrar, NASA's MTPE became the Earth Science Division. The name change was in line with Asrar's strategy to fly under the political radar of intensified climate change politics. The relative autonomy Goldin had granted Kennel and Townsend did not apply to Asrar, an able scientist but challenged bureaucratic operator and coalition builder. He was clearly given a very difficult hand of cards to play. Goldin was adamant that NASA not get into "operations," by which he meant repetitive monitoring rather than developing innovative technology.

This policy had major impacts on Asrar's tenure. One was Goldin's decision to cede the planned two follow-on sets of EOS satellites, or at least their instruments, to NOAA. This transfer had the effect of taking away a good part of NASA's Earth Science future. The same new-technology-first policy also led to a decision to make a future Jason-2 conform to the Administrator's innovation emphasis. There was a novel "wide-swath" concept that was to be employed to make Jason-2 better technically and more acceptable to the NASA Administrator. It proved too great a technical and financial stretch, however. A third impact was that Asrar was obliged to transfer Jason-2 to NOAA as soon as possible to make room for the "new." But what was desirable for NASA was not what NOAA wanted, particularly in terms of costs. Finally, all these moves had to be implemented at a time when the faster, better, cheaper mantra of Goldin put an emphasis on doing more with less NASA budget, particularly as Goldin emphasized Mars over Earth.

Asrar succeeded in getting satellites he inherited relevant to sea-level rise launched during his tenure. He maintained the program and sought a broader organizational underpinning via EUMETSAT. His struggle was with money for new starts to secure the future. His problems worsened under Goldin's successors, O'Keefe and Griffin.

O'Keefe gave Asrar support, but reorganized Earth Sciences as a subordinate element of an overall Science Directorate. Asrar became a deputy director of this Science Division, and his Earth science constituency saw this move as a demotion and Asrar as losing bureaucratic influence on which they depended for funds. Griffin further hurt Asrar's cause when he made careless comments about climate change being a problem with which NASA did not need to deal. Finally, Asrar could not continue the Kennel strategy of flying under the climate politics radar. NASA's most famous climate scientist, Hansen, regarded those who did not sound the alarm on climate change as guilty of "scientific reticence" in the face of disastrous reality, particularly owing to sea-level rise.

Asrar kept his program going, but lost the support of a core constituency, namely the Earth scientists. This community turned away from Asrar and his superior, Griffin, to Congress and the National Research Council to avoid what it called a "collapse" of the Earth Science satellite program.

The Rebuilder and Institutionalizer: Michael Freilich, 2006–2019

Freilich's 13-year tenure as head of Earth Science was long, eventful, and positive. It had two overlapping phases. The first was the rebuilding and reinvigorating of Earth Science, largely from 2006 to 2016. The second embraced the 2016–2019 period when institutionalization took place. What Freilich had to do in phase 1 was to incrementally augment his program's budget in line with a plan for projects based on the National Research Council's Decadal Survey. To do that, he had not only to fortify a technical program, but the political base for it and do so in an environment of divisive partisan conflict.

Freilich supported all aspects of the Earth Science program. But, as an oceanographer, he was personally attentive to sea-level rise and its causes. Fully aware of the sea-level threat, as a believer in climate change, he continued and strengthened an integrated program of ocean satellites and glacial-melting research, especially in Greenland. Intensely mission-oriented, he felt a deep sense of calling and responsibility that helped him enthuse and lead others in the United States and abroad. He believed bureaucracy's benefits outweighed its limits by achieving collective action for large-scale needs.

Like all his predecessors, he subscribed to Earth System Science as a long-term visionary goal and adopted an incremental and distributive satellite model to achieve it. Integration was the key, and sea-level rise stood out in showing how integration could be achieved. He kept "data continuity" flowing in spite of a gap in key satellites through activities like the airplane-based ICEBRIDGE. He was an astute coalition-builder who forged alliances over long years of effort. He followed his own star, based on the NRC Decadal Survey he inherited when he arrived. That stance brought him into conflict at times with political masters from both parties.

Although determined to protect his science emphasis, he bent to reality and societal need in the operational direction he had resisted for years. Uncertain of NOAA, European officials turned to Freilich to lead in keeping the transatlantic program of continual observation going when it threatened to unravel. He was an advocate for collaboration in the interest of not only NASA, but the planet. Sentinel 6/Jason CS secured data continuity throughout the 2020s and beyond. At the same time, the NASA–ESA alliance protected NASA's core mission of science. ESA could carry the load of operations, or so Freilich hoped.

The test of institutionalization for the climate change/sea-level mission was its survival under Trump. Freilich's long tenure and skills allowed him to negotiate with OMB and Congress to protect his program during a decidedly unfavorable political environment. He knew how to maneuver in Washington, collaborate internationally, and make bureaucracy work. He left NASA with his program intact and with ambitious plans for the future based on a second Decadal Survey.

The launch of a Sentinel satellite named for him in 2020, and a new generation satellite, SWOT, in 2022, demonstrated that the program he had led for 13 years was ready for another chapter.

Toward the Future

A unifying goal, one encompassing aspirational and "working" elements, supportive constituency, and relay of bureaucratic leaders with varying but sufficient coalitional skills have made NASA's climate change/sea-level rise program a success. These leaders, mainly mid-level science administrators ("technocrats"), were ultimately effective in spite of what was often a dysfunctional political environment. Program building is never easy, but it is especially so when the program is controversial. Science is not immune from the vagaries of politics.

At the time of writing, Karen St. Germain, who became head of Earth Science in 2020, had the task of taking this program to the next step. She gained a highly supportive political environment thanks to President Joe Biden in 2021. He made climate change a top priority. That meant a rise in budget, well beyond the $2 billion mark for Earth Science. Also, there was growing acceptance by politicians and the public of climate change and need for dealing with its impacts, albeit not with the urgency required. NASA was expected to play a significant role, including more active outreach in the interest of making science usable.

St. Germain likely has years to make a record. She benefits from the guidance of the Decadal Survey and the momentum of the Earth System Observatory decision. Her task is to be an augmenter of an established program.

What this book shows is that government can work. The process is messy and conflicted. But where there is vision united with human ability, bureaucratic agencies and their allies can compensate for macro-political divisions. To be sure, bureaucracy-driven policy is slow, uneven, and sometimes overly cautious. Everyone laments "bureaucratic inertia." But the other side of inertia is the merit of persistence. NASA did persist over decades, pulling together a technological system for measuring and understanding a truly serious problem. Building that satellite/research system took decades. The process was incremental and painfully slow. But in the end, it was accomplished.

The governmental challenge for the future is to improve and apply this system to make wiser national and international policy for adaptation and mitigation. Climate change and sea-level rise are not going away. NASA, as one powerful engine behind science and technology, can help. To function best, that engine needs a more coherent and comprehensive policy context, one that fills gaps between knowledge and action and provides long-term support. What NASA and bureaucracy-driven policy generally can provide is staying power and forward movement in the midst of political turbulence. That will be abundantly required in the years ahead.

Index[1]

[1] Note: Page numbers followed by 'n' refer to notes.

W. H. Lambright, *NASA and the Politics of Climate Research*, Palgrave Studies in the History of Science and Technology, https://doi.org/10.1007/978-3-031-40363-7

C

D

E

S

T

U